Modus Ponens (MP)
One may infer from two previously obtained sentences of the forms

$$p \supset q$$

and

$$p$$

the corresponding sentence

$$q.$$

Modus Tollens (MT)
One may infer from two previously obtained sentences of the forms

$$p \supset q$$

and

$$\sim q$$

the corresponding sentence of the form

$$\sim p.$$

Hypothetical Syllogism (HS)
One may infer from two previously obtained sentences of the forms

$$p \supset q$$

and

$$q \supset r$$

the corresponding sentence of the form

$$p \supset r.$$

Constructive Dilemma (CD)
One may infer from three previously obtained sentences of the forms

$$p \vee q$$
$$p \supset r$$

and

$$q \supset s$$

the corresponding sentence of the form

$$r \vee s.$$

Rules of Replacement for '⊃' and '≡'

Transposition (Trans)
$$p \supset q :: \sim q \supset \sim p$$

Implication (Impl)
$$p \supset q :: \sim p \vee q$$

Exportation (Exp)
$$(p \;\&\; q) \supset r :: p \supset (q \supset r)$$

Equivalence (Equiv)
$$p \equiv q :: (p \supset q) \;\&\; (q \supset p)$$
$$p \equiv q :: (p \;\&\; q) \vee (\sim p \;\&\; \sim q).^{6}$$

(Continued inside back cover)

William Gustason
Dolph E. Ulrich

Purdue University

ELEMENTARY SYMBOLIC LOGIC

Holt, Rinehart and Winston, Inc.

*New York   Chicago   San Francisco   Atlanta   Dallas
Montreal   Toronto   London   Sydney*

To our parents

PREFACE

This textbook is designed for a one-semester or two-quarter introductory course in symbolic logic. No previous training is presupposed, and a deliberate attempt has been made to appeal to students from a wide range of disciplines, who enroll in such a course with varying motives and expectations. For the benefit of students with nontechnical backgrounds, relatively intricate topics—especially those pertaining to quantification—are discussed more extensively than is usual. The emphasis is thus on developing the student's grasp of standard techniques and concepts rather than on achieving the degree of sophistication and rigor that would be appropriate for a more specialized and demanding course.

In light of the plethora of logic textbooks available, the appearance of a new one calls for some explanation. Our chief aim has been to provide the instructor with considerable flexibility when choosing an order in which to present standard topics. Accordingly semantic methods (Chapter 2) and natural deduction (Chapter 3) for sentential logic are developed independently, the preliminary discussion of the sentential connectives in Chapter 1 providing a common lead-in to both. Thus the instructor may begin in the customary fashion with truth-tables and then proceed to natural deduction; alternatively, he may begin with deductions and follow up with truth-tables. In either case he may at his option finish off with Chapter 4, where the two approaches are linked by a brief discussion of appropriate metatheoretic questions. This book hence differs from the majority of tests, where one must begin with truth-tables, and from the remainder that require coverage of deductions first.

Moreover, the ten rules of inference and replacement for '&', 'v', and '~' presented in Chapter 3 are by themselves semantically complete, and this makes possible additional variations that may intrigue some instructors. With a little maneuvering it is possible to begin with truth-tables and deductions for just these three connectives, proceed to Chapter 4 for the completeness discussion, and then return to Chapters 2 and 3 to develop (in either order, of course) truth-tables and rules for '⊃' and '≡'. Thus, in addition to the two standard orders

Chapters 1, 2 (last two sections optional), 3, and 4 (optional)
Chapters 1, 3, 2 (last two sections optional), and 4 (optional)

one has on this third approach other options, for example:

Chapter 1 (omitting 1.4 temporarily), Section 2.1, 2.3 (first half), 3.1, Chapter 4, 1.4, balance of Chapters 2 and 3 (in either order).

In approaching quantification theory, similar options are available. After the discussion of symbolization and the associated apparatus in Sections 5.1 and 5.2, one may cover—in either order—semantics (5.3) or natural deduction (6.1 and 6.2, with 6.3 optional), returning eventually to the other. The sections on identity and definite descriptions (5.4 and 6.4) may be covered in their natural order or left till the end if desired; and of course they may be omitted should time run short. Sample orders of presentation include:

5.1–5.4, 6.1–6.4
5.1–5.3, 6.1–6.3, 5.4, 6.4
5.1, 5.2, 6.1–6.3, 5.3, 5.4, 6.4

The bulk of the text, then, is taken up with symbolization, semantics, and natural deduction for sentential and quantificational logic. But further material is available for times when some variety is desired or for background reading – a discussion of functional completeness in Section 2.5 for example, and of semantic completeness and soundness in Chapter 4. There is also an appendix introducing mathematical induction, which many will find a useful supplement to Chapter 4, and one on axiomatic systems. In addition, an appendix on normal forms and circuits is provided to augment Chapter 2, and another on truth trees for those who desire a more systematic method than that of Section 2.4. Any of these can be inserted at appropriate points – and all may be omitted – without affecting continuity or the coverage of essentials. Optional sections and certain exercises which might best be regarded as "projects" are marked with stars. We note that the first chapter, concerned mainly with preliminary terminology and notation, offers many temptations for squandering class time. Much of the material there can be mastered by the student largely on his own, however, and the authors have found that three or four class meetings are sufficient for covering its essentials.

We have tried to write for the *student*, showing him by example how to think about the problems and methods being treated. Particular emphasis is placed on translation into quantificational notation and strategy in discovering deductions. And since students often want to insert tautologies into deductions we have a rule allowing insertion of those of the form '$p \vee \sim p$'. There are surprising advantages – certain deductions turn out to be more easily constructed than in systems lacking such a rule, and conditional proof turns out to be redundant (though of course it is employed as a practical device). In addition, the strategy of indirect proof emerges in a very natural way. It should also be noted that the student who works through Chapter 4 will be able to *demonstrate* the redundancy of conditional proof by obtaining a deduction theorem for the system developed in Sections 3.1 and 3.2.

In the rules for quantificational logic we have tried to put into sharper focus the role often played by free variables. The letters 'u', 'v', 'u'', 'v'', etc., which cannot appear as bound variables, are made available particularly for use with the rules **EI** and **UG**, which consequently can be stated without contorted clauses concerning freedom and bondage. Some instructors may choose to reserve 'v', 'v'', etc., for use with the former and 'u', 'u'', etc., for the latter, though our rules do not require this, of course. Well-known semantic considerations have led some authors to cast **EI** as a sub-proof rule. Such formulations strike many students as nonintuitive and puzzling, however, so we have preferred what seems a more natural version. Certain arguments require circuitous deductions but their number can be kept to a minimum with judicious symbolization.

Anyone who writes a textbook of course owes many debts, conscious and unconscious, to other authors. We have acknowledged in footnotes the ones we are aware of, and here thank those whose influences have been less obvious to us. We would like also to thank our friends, former teachers and students. Professor Jack Nelson must be singled out for special thanks; he used a draft in his course at Temple University and made many valuable and detailed suggestions. Thanks go also to our colleague Professor William L. Rowe, and to Ken Tohinaka, who used drafts in their courses as well and suggested improvements. We are grateful to the publisher's reviewers for their contributions, and to John Tugman and Brian Heald of Holt, Rinehart and Winston. Our thanks extend, finally, to Marie Allison who typed an earlier draft of the manuscript, to Ralph Moon for many suggestions and for assisting in preparing the index, and to Susan Ulrich, who helped in countless ways from the time of the book's inception until its publication.

West Lafayette, Indiana   W. G.
                                                                                                              D. U.

CONTENTS

CHAPTER ONE

INTRODUCTION

1.1 Arguments

Like most disciplines, logic treats so wide a range of problems and topics that any attempt to describe its subject matter briefly is bound to be misleading and incomplete. But its most elementary and fundamental concern is with matters relevant to the study of arguments — the primary topic of this book. For our purposes an **argument** may be characterized as a sequence of sentences of which one — the **conclusion** of the argument — is marked off as *following from* the others — which are the **premises** of the argument. Our main job will be to develop some tools and techniques useful in determining whether or not certain sentences follow from certain others, and in explaining why they do or do not.

A number of difficult but interesting and important questions are partly questions about what follows from what; for example:

Do there exist relatively simple physical laws from which all other physical laws follow?

Does there exist a series of relatively simple arithmetic truths from which all other arithmetic truths can be deduced?

Do predictions about what is going to happen tomorrow follow from certain statements about what has happened in the past?

May we infer from certain things we all know, or can readily establish, that there exists a Supreme Being whom we may call 'God'?

Because of the intrinsic complexity of such questions, however — not to mention the highly specialized knowledge of other disciplines that most of them call for — we shall generally be dealing with considerably simpler ones and as a result with arguments that are somewhat less interesting.

We need not resign ourselves, though, to a steady diet of banalities like the classic

All men are mortal.
Socrates is a man.
Therefore Socrates is mortal.

We shall find use for such arguments, especially in the early sections of several chapters where it will be helpful to begin with such straightforward examples. But we shall also get a chance to deal with a number of more interesting arguments, some of which touch — as do certain of the following examples — on such questions as those mentioned above.

As a first step, the reader is asked to examine the following paragraphs critically, asking himself which of the arguments presented therein are good and which bad. At the same time, he should give the reasons on which he bases his decisions.

(I) If God can create a stone too heavy for Him to lift, there is something He cannot do; and if He cannot create a stone too heavy for Him to lift, there is something He cannot create. If there's something God cannot do He is not omnipotent, and if there's something He cannot create He is not omnipotent. Therefore God is not omnipotent.

(II) If it is morally permissible for me to perform a certain action, then it must be morally permissible for everyone to perform that action. But if the consequences of everyone's performing a certain action are on the whole undesirable, then it is not morally permissible for anyone to perform that action. Therefore, if the consequences of my performing a certain action are, on the whole, undesirable, then it is not morally permissible for me to perform that action.

(III) If we introduce the Arabic notation for numbers, and pile a decimal notation on top of that, we sometimes find ourselves blessed with two different ways of writing names for one and the same number. For example, the number we denote by a vertical stroke (and call 'the number one') is identical with the number whose name is sometimes written '.999 $\cdots$'. For let $n = .999 \cdots$ Then $10 \times n = 10 \times .999 \cdots$, that is, $10 \times n = 9.999 \cdots$ But then $(10 \times n) - n = 9.999 \cdots - n$, so that $9 \times n = 9.999 \cdots - .999 \cdots$ Consequently $9 \times n = 9$, that is, $n = 1$. But $n = .999 \cdots$ and so $1 = .999 \cdots$, as was to be shown.

(IV) If freely falling bodies fall faster in proportion to their weight, then a three-pound body would fall faster than one of two pounds. But suppose the three-pound body consists of a two-pound and a one-pound body fastened together. Then it would fall more *slowly* than a single two-pound body, for the one-pound body would retard the motion of the attached two-pound body. Therefore, freely falling bodies do not fall faster in proportion to their weight.

(V) One Saturday, a prisoner was sentenced to be hanged. "The hanging will take place at 8:00 AM ," the judge told him, "on Monday, Tuesday, Wednesday, Thursday, or Friday of next week. But you will not have any idea which day it will be until you are so informed at dawn on that day."

Alone later in his cell, the prisoner began to think about the judge's remarks. Obviously they could not wait until the *last* day to hang him, without violating the judge's decree. "If the first four mornings went past and I were still alive," he thought to himself, "I'd know for sure by that fourth afternoon when the hanging would be—there would only be one day left when it *could* be! But the judge said that I *would not* be able to tell in advance like that. So the *last* day is absolutely ruled out as the day for the hanging; it can only take place on Monday, Tuesday, Wednesday or Thursday." So Thursday, he knew, was the last day they could pick for the hanging. But of course they could not wait until the *last* day .... "Thursday's not a real possibility either, then," thought the prisoner. "Were I still alive at dusk on Wednesday, I would know that Thursday had to be the day. Friday has already been ruled out.

"So Wednesday seems to be the latest day the judge could pick. But for the same reasons, Wednesday can be eliminated as a real possibility ... and then Tuesday, and finally Monday. They cannot possibly hang me in accordance with the judge's decree!"

Does the reader need to be told that the prisoner was quite surprised when the man in the black hood came to call on Wednesday, just at dawn?

(VI) It is sometimes supposed that (declarative) *sentences* are the things that may properly be said to be 'true' or 'false'—when one says, for example, that it is true that snow is white, it is the sentence "Snow is white" that is being said to be true.

Suppose Downs, in Detroit one bright morning, looks quickly out his window on which a lawn sprinkler is playing, erroneously concludes that it is raining, and utters to his wife the sentence "It is raining," meaning thereby to assert that it is raining that morning in Detroit. And suppose that Smith in Seattle that same morning sees a genuine downpour out his window and utters to his wife the sentence "It is raining," meaning thereby to assert that it is raining that morning in Seattle. Then what Downs says is false, and what Smith says is not.

When someone utters a sentence, and thereby succeeds in asserting something, at least *two* things get done: first, a sentence is uttered, and second, something is said (claimed, asserted) that is either true or false. The considerations advanced in the preceding paragraph suggest that we should not identify what is said, which is either true or false, with the *sentence* uttered in the saying of it.

Now if the reader reflects on whatever objections he may have raised to these arguments, he will very likely find that they fall into two groups. In the one group are those objections involving a claim that one or more of the premises of a certain argument are false; in the other are those involving the claim that the conclusion of the argument in question simply does not follow from its premises. The distinction between these two sorts of criticism to which an argument may be subjected can be drawn more easily if we look at four comparatively trivial examples, where a more exhaustive discussion is feasible.

(VII)   The cube of four is either 16 or 64.
It is not the case that the cube of four is 16.
Therefore the cube of four is 64.

(VIII)   Either all the premises of argument (VII) above are true or else some of the premises of argument (VII) are false.
Not all the premises of argument (VII) are true.
So some of the premises of argument (VII) are false.

(IX)   Either litmus turns blue in a basic solution or else litmus turns blue in an acid solution.
Litmus turns blue in a basic solution.
Therefore litmus turns blue in an acid solution.

(X)   Either the moon is larger than the earth or else the moon is larger than the sun.
The moon is larger than the earth.
Therefore the moon is larger than the sun.

Which of these arguments are good arguments, and which are bad ones? If asked to label them, most would probably write

(VII)   Good

(VIII)   Bad

(IX)   Bad

(X)   Bad

though some might put 'Good' beside (VIII) instead.

(VII) is pretty clearly a good argument; its premises are both true, and its conclusion certainly follows from them. (X) is just as clearly bad. Its second premise is false; its first premise is also false; and even if both premises were true, the conclusion still would not have to be, for it does not follow from them.

And now we can notice that (IX) suffers from one of the defects of (X) and (VIII) from the other. What's wrong with (IX) is, to put it simply, that its conclusion just does not follow from its premises. (We may not yet be able to *explain* exactly what we mean when we say that a particular sentence does not follow from certain others, but in such an easy case as this one we can certainly *recognize* what's going on.) Incidentally, this is just what's right about (VIII), and what probably explains why some labeled (VIII) 'Good' rather than 'Bad'. Argument (VIII) is bad in one sense, because its second premise is not true. But if the premises of (VIII) *were* both true, the truth of its conclusion would then be guaranteed — the conclusion of (VIII) certainly does follow from its premises.

All this suggests a way for us to improve a little on our vague 'Good'– 'Bad' terminology. Why not check each of our arguments from *two* standpoints? First ask: *are all its premises true*? Then ask: *regardless* of whether or not the premises of this argument are all true, *does its conclusion follow* from them?

(VII) Good
All premises are true
Conclusion follows from them

(VIII) Bad
Not all premises true
Conclusion does follow from them, though

(IX) Bad
All premises are true
Conclusion, however, does not follow from them

(X) Bad
Not all premises true
Conclusion does not follow from them — even if they *were* all true, the conclusion wouldn't have to be

It is customary to use some special terms in discussing these matters. An argument is said to be **valid** if its conclusion follows from its premises, and **invalid** if its conclusion does not follow. Alternatively, we shall sometimes say that the premises **imply** the conclusion when the latter follows, and that they do *not* imply it when it does not follow. Thus, every valid argument with true premises must have a true conclusion, but an invalid argument can have true premises and yet a false conclusion. When an argument is valid *and* all its premises are true, it is **sound**.

Arguments which are not sound—arguments, that is, which either are invalid or have at least one false premise—are called **unsound**.

Returning to the old examples with our new terminology, we have:

(VII)	Good	All premises are true Conclusion follows from them	Sound
(VIII)	Bad	Not all premises true Conclusion does follow from them, though	Unsound (though valid)
(IX)	Bad	All premises are true Conclusion, however, does not follow from them	Unsound (because invalid)
(X)	Bad	Not all premises true Conclusion does not follow from them—even if they *were* all true, the conclusion wouldn't have to be	Unsound (on *both* counts)

The terms 'good' and 'bad', which were never very useful ones, may now be discarded. If someone gives an argument whose conclusion we think is suspect, there are really only two legitimate ways we can attack him. Either we must challenge one or more of his premises, or else we must claim that his conclusion does not follow from them. If we can do neither of these things—if we admit that all the premises of his argument are true, and also admit that the conclusion follows from them—then of course we must admit that the conclusion is true.

Of course, the premises of different arguments concern topics in a wide variety of disciplines, and as a result the job of checking them for truth or falsity falls most often within the province of a specialist in some area other than logic—a philosopher, a chemist, a mathematician, a literary scholar, or the like. It is clearly not the business of logic to dip litmus paper into lye, for example, nor to measure the circumference of the moon. Our concern in the remainder of the test will not be with soundness and unsoundness, then, but with validity and invalidity—features of an argument which will be seen to be independent of the specific content of the sentences comprising it.

DEFINITION An argument is **valid** (its premises **imply** the conclusion) if and only if its conclusion follows from its premises; that is, if and only if it is impossible for all its premises to be true but its conclusion false.

DEFINITION   An argument is **invalid** (its premises do not imply the conclusion) if and only if its conclusion does not follow from its premises; that is, if and only if it is at least possible for all its premises to be true but its conclusion false.

The usefulness of these definitions is limited, of course, by the vagueness of such terms as 'follow from', 'not follow from', 'possible', and 'impossible'. It will be the task of the rest of this book to offer some analyses of these terms as they might be applied to a wide range of arguments found in everyday and technical discourse.

For now, we only note that the terms 'possible' and 'impossible' must be sharply distinguished from the terms 'probable' and '(highly) improbable' — a distinction paralleling that between **deductive** methods for evaluating arguments and **inductive** methods. Only the first of these methods will be developed in this book. We mention inductive methods merely to contrast the two. The following arguments, for example, are most naturally assessed from an inductive standpoint;

(XI)   Jones has hated Smith for years.
Jones badly needed money to pay off a gambling debt.
Jones was having an affair with Smith's wife, who would collect on Smith's insurance policy in case of his death.
Smith was shot with a .38 caliber pistol belonging to Jones.
Two reliable witnesses saw Jones leaving Smith's house on the night of Smith's murder about fifteen minutes after the coroner's estimated time of death.
Jones's fingerprints were found on the murder weapon.
Smith's wife testifies that she conspired with Jones to murder her husband.
Therefore Jones murdered Smith.

(XII)   Jones has hated Smith for years.
Jones badly needed money to pay off a gambling debt.
Jones was having an affair with Smith's wife, who would collect on Smith's insurance policy in case of his death.
Smith was shot with a .38 caliber pistol belonging to Jones.
Two reliable witnesses saw Jones leaving Smith's house on the night of Smith's murder about fifteen minutes after the coroner's estimated time of death.
No fingerprints were found on the murder weapon.
Smith's wife admits an affair with Jones, but denies any murder plot.
Therefore Jones murdered Smith.

In neither case does the conclusion follow from the premises. It is at least *possible*, albeit not *probable* nor perhaps even very likely, for them all to be true without Jones having murdered Smith. From the point of view of deductive logic, then, each is invalid and there is little more to be said.

But the methods of inductive logic employ different criteria in investigating the relation between the premises and conclusions of such arguments. The assessment is made according to the relative degree of *evidence* the premises provide for the conclusion — the greater the support or evidence, the stronger would the argument be said to be. Depending on the extent to which the premises confirm or disconfirm the conclusion, we would say that the conclusion is *probable* or *improbable* with respect to them. When an argument is assessed as being *strong*, there is a high probability that its conclusion is true, and when assessed as *weak* its probability is low (relative to the evidence supplied by the premises in both cases).

Deductive evaluations are all-or-nothing affairs. When an argument is valid, it is *impossible* for its premises to be true while its conclusion is false. In contrast, inductive evaluations are matters of degree, ranging from very low to very high. When an argument is strong it is *highly improbable* that its premises are true but its conclusion false. The degree would presumably be high in the case of argument (XI) above (perhaps to the extent that a jury would consider the conclusion to follow *beyond reasonable doubt*), fairly low in the case of argument (XII) (the evidence is slim, and mostly circumstantial).

Inductive inference is very common in science, law, and everyday life; for that reason, it is no less important a topic for study than deductive inference. We omit further discussion of inductive inference hereafter mainly because a thorough and systematic study of it is best undertaken only after a familiarity with deductive methods has been acquired.

EXERCISES

    I. Construct a brief, clear-cut example of your own of:
1. an invalid argument whose premises are all true and whose conclusion is true as well;
2. an invalid argument whose premises are all true but whose conclusion is false;
3. an invalid argument whose premises are all false and whose conclusion is true;
4. an invalid argument whose premises and conclusion are all false;
5. a valid argument whose premises are all false but whose conclusion is true;

6. a valid argument whose premises are all false and whose conclusion is false as well;
7. a valid argument whose premises are all true.

II. Explain why any correct answer to (7) above consists of an argument whose conclusion is true.

*III. Construct arguments of your own concerned with:
1. the existence or nonexistence of numbers with some property or properties you regard as interesting;
2. the existence or nonexistence of a machine capable of thinking;
3. the existence or nonexistence of human beings other than yourself;
4. the speed of light;
5. the effect of a species' environment on its development over the centuries;
6. the effect of a human being's environment on his cultural development.

Which of your arguments are valid and which invalid? Which are sound and which unsound? Which might be more fruitfully studied from the point of view of *inductive* logic?

1.2 'And', 'Or', and 'Not'

Arguments are our prime concern, but because they are themselves sequences of *sentences* we must begin by looking at the latter. We intend to restrict our attention to those contexts in which each sentence that occurs is either true or false. In so restricting ourselves we do not of course mean to suggest that every sentence is either true or false—imperatives and interrogatives, for example, are clearly neither—nor to deny that a sentence may be used in different contexts to say different things. But the tools and techniques to be developed in this book are only intended to be applicable in those contexts in which each sentence encountered is used unambiguously to say something either true or false.

Strictly speaking, considerations similar to those advanced in example (VI) of the preceding section suggest that it may be improper to apply the terms 'true' and 'false' to *sentences*. Suppose, however, that we introduce a technical term and speak instead of the **truth-value** of a sentence in a particular context (provided that the sentence is used unambiguously in that context and, in all its occurrences, is used to make just one claim). A sentence will be said to **have the truth-value T** in a particular context if and only if it is used unambiguously there and expresses something *true*, and will be said to **have the truth-value F** in that context if and only if it is used unambiguously and expresses something *false*. We can then restore a somewhat common way of speaking: for the sake of *convenience* only, we can agree to call a sentence with the truth-value **T** in a particular context 'true' (in that context), and one with the truth-value **F** in that context 'false' (in that context). Nothing in our manner of speaking has changed, but everything is different; when *precision* is more important than convenience, such talk can always be replaced by talk about the truth or falsity of that which, in that context, the sentence expresses.[1]

It will be convenient also to borrow a pair of terms from the grammarians, though our definitions will differ slightly from theirs, and distinguish **compound sentences** from **simple sentences**. The former contain *other* sentences—more precisely, other independent clauses—as proper parts; the latter do not. Thus 'Socrates is wise and Plato is wise', 'Either Socrates is wise or Plato is wise', and 'It is not the case that Socrates is wise' are all compound sentences since each contains at least one other

---

[1]Depending upon the reader's philosophical predilection, these things that sentences may be used to express which are true or false in the primary sense may be thought of as statements, propositions, or whatever. Readers without such predilections may acquire some by reading P. F. Strawson, *Introduction to Logical Theory*, London: Methuen, 1952, Chapters 1 and 6; or W. V. Quine, *Philosophy of Logic*, Englewood Cliffs: Prentice-Hall, 1970, Chapter 1. See also R. L. Cartwright, "Propositions," in R. Butler (Ed.), *Analytical Philosophy*, First Series, New York: Barnes and Noble, 1966.

sentence — 'Socrates is wise' or 'Plato is wise' — as a proper part. Since the latter contain no other sentences, however, they themselves are simple.

Simple sentences can be regarded, then, as the building blocks from which compound sentences are constructed, and such expressions as 'and', 'either ... or ---', and 'it is not the case that ...' may be thought of as the mortar used to hold them together. These three expressions will be of particular interest to us because our initial concern will be with compound sentences that contain them and whose truth-values are determined solely by the truth-values of their component sentences. Such sentences suffice for the construction of a number of easily analyzed arguments that serve as our starting points in the next two chapters.

When one sentence can be obtained from another by prefixing 'it is not the case that ...' to it, the former is called the **negation** of the latter. Thus

(1) It is not the case that God is omnipotent

is the negation of

(2) God is omnipotent;

and

(3) It is not the case that it is not the case that God is omnipotent

is, in turn, the negation of (1).

If two sentences are so related that they must have opposite truth-values, they are said to be **contradictories**, and each is a **contradictory** of the other. (1) and (2) are contradictories, then, as are (1) and (3). Indeed, given our restriction to sentences that are either true or false, an important general point emerges: a sentence and its negation will always be contradictories. Where 'p' represents any sentence, its negation — the corresponding sentence of the form 'it is not the case that p' — will be *false* if 'p' is true and *true* if 'p' is false.

In many cases, of course, English allows us to get the effect of prefixing 'it is not the case that ...' to a sentence without going to quite so much trouble. Sometimes we can simply prefix 'not' to the sentence ('*Not* all men are mortal') or insert it into the verb ('God is *not* omnipotent'). Certain verbs require 'does not' instead ('Litmus *does not* turn blue in an acid solution') and still other variations are possible (such as 'Jim is *un*married', 'Argument (VII) is *in*valid', 'Hercules is *im*mortal', and 'Smedley is *no* ladies' man'). It is natural to extend our terminology slightly and call these sentences 'negations' as well. But one should not presume that such options are always open to us. If we try inserting 'not' into the verb of 'Some dogs are spaniels', for instance, we produce 'Some dogs are not spaniels', and this clearly does not have the same force as 'It is not the case that some dogs are spaniels'. Indeed, since 'Some dogs are spaniels' and 'Some dogs are not spaniels' are *both* true, they are not contradictories and so neither could be the negation of the other.

Suppose now that we begin with (2) and

(4) God is benevolent

and combine them into a single compound sentence by inserting the word 'and' between them:

(5) God is omnipotent and God is benevolent.

Clearly *both* (2) and (4) must be true for (5) to be true — if God is not omnipotent, for example, or if He is not benevolent, (5) is false. Compound sentences like (5), which would be true if both components were true but false otherwise, are called **conjunctions** and their components are called their **conjuncts**.

Again, English syntax provides alternative constructions. In neither

(6) God is omnipotent and benevolent

nor

(7) God and UNIVAC are omnipotent,

for example, does 'and' serve to connect two sentences. But it is clear that (6) will be true if and only if (2) and (4) both are, and that (7) will be true if and only if (2) and 'UNIVAC is omnipotent' are, so we will count them as conjunctions as well.

Of course we might also combine (2) and (4) into a single sentence by inserting the word 'either' before the first and 'or' between them:

(8) Either God is omnipotent or God is benevolent.

(As with negations and conjunctions, alternate forms of speech are common, such as 'God is omnipotent or benevolent'.) When we try to state the conditions under which (8) would be true, however, a complication arises. It is clear that it will be true if just one of its components is true and false if neither of them is true, but what about the case where *both* components are true? What if God is not only omnipotent but benevolent as well?

In most contexts a person who uttered (8) might justifiably say that in doing so he was only claiming that *at least* one of its components is true because he was using 'either ... or ---' in the sense of 'either ... or --- or both'. In such a case we would surely count (8) as true. This is, of course, a very common way of using 'either ... or ---'; a sentence in which it is used in this way will be called a **nonexclusive disjunction**,[2] and its components its **disjuncts**. In almost any context one can imagine, the sentence "All who are over 65 or who require major surgery will receive hospital benefits" would involve a clear case of nonexclusive disjunction, for the intent would surely be that a person who was *both* over 65 *and* required major surgery would receive benefits just as would anyone meeting only one of those conditions. And one who says "The senator's remarks reveal that he is either stupid or ignorant of the issues" does not mean to rule out the possibility that those remarks result from both stupidity and ignorance.

On the other hand one can imagine a different context in which (8) was uttered in order to sternly quash the supposition that God was both omnipotent and benevolent. An age-old argument, for instance, contends that the two qualities are incompatible in a Supreme Being. Were He benevolent He would want a perfect universe and were He all-powerful He could have one; since the universe is not perfect — so the argument goes — He clearly cannot be *both* omnipotent and benevolent. In the context of such a debate (8) would perhaps have to be understood as synonymous with a sentence of the form 'either ... or --- *but not both*'. Sentences of this sort will be termed **exclusive disjunctions**, and they are clearly false when both their *disjuncts* are true. Examples of exclusive disjunctions seem harder to come by than examples of nonexclusive ones, but think of a weary husband who has been on a day-long shopping trip with his wife. In response to her remark that they still must buy groceries and visit her relatives he might say in exasperation "*Either* we buy groceries *or* we visit your relatives." He obviously does not intend to do both, and if both are done he will have to add the utterance of a false sentence to his woes.

---

[2]That this term is more awkward than the standard 'inclusive disjunction' is not really a consideration, since it shortly gives way to the term 'disjunction'.

To determine in context the truth-value of a sentence containing 'either ... or ---', then, it is sometimes necessary first to determine whether it is being used in the sense of 'either ... or --- (*or both*)' or in the sense of 'either ... or --- (*but not both*)'. In the first case the sentence would be a nonexclusive disjunction and so would be true if at least one disjunct were true and false only if both disjuncts were false. In the latter case, however, it would be an exclusive disjunction and so would be true if and only if its disjuncts differed in truth-value and false if their truth-values coincided. Both kinds of disjunction are false when both disjuncts are false and true when just one disjunct is true, but nonexclusive disjunctions are true — and exclusive ones false — when both disjuncts are true.

Of course it often makes no difference. In such examples as

Either $2 + 2 = 4$ or $2 + 2 = 5$
Either Jones is now in New York or he is now in Paris
Either Simmons is a bachelor or he is married

the component disjuncts are incompatible — they could not both be true. It would be a mistake, though, to conclude that these examples must be construed as exclusive disjunctions. The fact that each could be followed by the phrase 'but not both' without any change in truth-value need not at all be the result of using 'either ... or ---' in the exclusive sense, but rather the result of disjoining two sentences whose incompatibility precludes any question of joint truth — a situation entirely unlike the remark of the weary husband. Indeed, these sentences may be interpreted in either sense — their disjuncts by themselves exclude one another, and it makes no difference to the truth-values of these disjunctions whether we read the 'either ... or ---' as reinforcing this exclusion or not.[3]

We've looked at three kinds of compound sentences: negations, formed typically by prefixing 'it is not the case that ...' to a sentence; conjunctions, formed often by inserting 'and' between two sentences; and disjunctions, formed by wrapping 'either ... or ---' around a pair of sentences. It is customary to call such compounding expressions **sentential connectives**, and important to notice that the truth-value of a compound sentence built up with the aid of the particular connectives we've discussed can often be determined once we know the truth-values of its components. Consider, for example,

(9) Either it is not the case that all human beings are mortal or it is not the case that both all human beings are men and all men are mortal.

---

[3]Cf. W. V. Quine, *Methods of Logic* (ed. 3), New York: Holt, Rinehart and Winston, 1972, pp. 11–16.

Since 'All human beings are mortal' is true, its negation — the first disjunct of (9) — is false. Since 'All human beings are men' is false and 'All men are mortal' true, their conjunction is false. So the negation of that conjunction — which is the second disjunct of (9) — is true. So (9) has exactly one true disjunct and it is true whether construed exclusively or nonexclusively.[4]

EXERCISES

I.   If we replace '*p*' with a true sentence, '*q*' with a false one, and '*r*' with a sentence whose truth-value is unknown to us, which of the following would be true? Which false? Which cannot be determined?

1.   Either *p* or both *q* and *r*
2.   *p* and either not-*q* or *r*
3.   Either not-*p* and *q* or *p* and not-*r*
4.   Either not-*p* or *q* and either not-*q* or both *r* and *p*

II. What if we replace '*p*' and '*r*' with false sentences, and '*q*' with a true one?

---

[4]From now on, we shall use the simple term 'disjunction' to mean 'nonexclusive disjunction'. When we want to talk about exclusive disjunctions, we shall use the term 'exclusive disjunction'.

1.3 '&', 'v' and '∼'

We now introduce some abbreviatory devices. To represent *simple* sentences we shall use capital letters of the English alphabet, calling them **sentence letters** to remind ourselves of their function. Using '*O*' for 'God is omnipotent' and '*B*' for 'God is benevolent', for example, the conjunction

(5) God is omnipotent and God is benevolent

can be shortened to '*O* and *B*'. It will be useful also to introduce some symbols to serve as *sentential connectives* — symbols that perform the same jobs as 'and', 'or', and 'not' by combining with sentence letters to produce compound sentences (or for short, just *sentences*). In place of the word 'and' in (5) it is natural, for example, to use the **ampersand**, '&', abbreviating (5) most fully as

(5′) *O* & *B*.

Since (5′) abbreviates a conjunction, we shall extend our terminology slightly and regard (5′) itself as a conjunction, and '*O*' and '*B*' as its conjuncts. If '*K*' represents a third sentence (say, 'God is omniscient'), '(*O* & *B*) & *K*' would also be a conjunction, with '*O* & *B*' and '*K*' as conjuncts. In such a case it is sometimes convenient to treat '*O*' and '*B*' as conjuncts of '(*O* & *B*) & *K*' as well.

For 'either ... or ---' in the *nonexclusive* sense — that is, for 'either ... or --- (*or both*)' — we will use the **wedge** symbol, 'v'.[5] Construed nonexclusively, then,

(8) God is omnipotent or God is benevolent

can be written

(8′) *O* v *B*.

Like (8), which it abbreviates, (8′) is a nonexclusive disjunction — or for short just a disjunction — and the expressions flanking the 'v' its disjuncts.

---

[5]Unlike English, Latin had two distinct words for 'or': *vel*, used for nonexclusive disjunctions, and *aut*, used for exclusive ones. Our symbol 'v' is derived from the first letter of the former.

As in the case of conjunctions, we will regard '*O*' and '*B*' individually as disjuncts of '(*O* v *B*) v *K*', along with '*O* v *B*' and '*K*'. As will be seen shortly, no special symbol for 'either ... or ---' in the exclusive sense will be needed.

The sign '~', called the **curl** or the **tilde**, will abbreviate the phrase 'it is not the case that'. Thus

(1) It is not the case that God is omnipotent

can be abbreviated

(1') ~ *O*,

and "It is not the case that it is not the case that God is omnipotent" can be written '~ ~ *O*'.

Parentheses must be included in our notation so that we can distinguish, for example, between

(10) ~ (*O* & *B*)

and

(11) ~ *O* & *B*.

(10) goes into English most idiomatically as 'God is not both omnipotent and benevolent', while (11) could be translated 'God is not omnipotent and He is benevolent'. Clearly, denying that God is *both* omnipotent and benevolent is not at all the same as affirming that he lacks the first of these attributes but possesses the second. Were God neither omnipotent nor benevolent, then (10), being the negation of a false conjunction, would be true, but (11), a conjunction with a false right conjunct, would be false.

In addition to sentence letters, sentential connectives, and parentheses, it will be convenient to use small letters from the middle of the alphabet — '*p*', '*q*', '*r*', '*s*' (adding, if the need arises, numerical subscripts as in 'p_1' and 'q_3') — as **sentential variables**. Unlike sentence letters, sentential variables will not be used to represent particular sentences but rather may be thought of as standing in a given content for *any* arbitrarily selected sentence. In the expression '*p* v *q*', for example, we do not have an abbreviation for a particular disjunction but something more like a schematic picture of disjunctions generally. In effect, '*p* v *q*' presents us with an overall *pattern* or *form* shared by many particular sentences; for this reason expressions like '*p* ∨ *q*' containing sentential variables are called **sentence forms**.[6] Sentence forms are not themselves sentences, and they do not have truth-values; rather they are employed to represent indiscriminately all sentences of the form depicted.

The sentences '*O* ∨ *B*', '*O* ∨ ~*K*', and '(*O* & *B*) ∨ *K*' are all derivable from '*p* ∨ *q*' by making appropriate substitutions for '*p*' and '*q*'. Similarly, the third of these sentences — but not the first two — is also derivable by substitution from '(*p* & *q*) ∨ *r*', and the second — but neither the first nor the third — is derivable from '*p* ∨ ~*q*'. It is helpful to remember that a sentence form like '*p* ∨ *q*' stands to such sentences as '*O* ∨ *B*' and '(*O* & *B*) ∨ *K*' as '*x* + *y*' stands to '3 + 1' and '(3 − 1) + 2' in arithmetic. It is precisely because an expression like '*x* + *y*' signifies no particular number that it is useful in discussing general properties shared by all numbers; and it is because sentence forms stand for no particular sentences that they will be useful in discussing features common to *all* conjunctions or to *all* disjunctions, and so forth.

Reflection on our earlier discussion of the phrases 'and', 'either ... or --- (or both)', and 'it is not the case that' enables us to define in their contexts of use our three sentential connectives '&', '∨', and '~' by specifying the conditions under which sentences containing them will be true or false.

Conjunction   A sentence of the form '*p* & *q*' is true if and only if both conjuncts are true, and false otherwise.

---

[6]More precisely, if less lucidly, we define the notion of a **sentence form** recursively as follows:

   i.   Each sentential variable ('*p*', '*q*', '*r*', '*s*', '*p₁*', '*p₂*', etc.) is a sentence form;

   ii. the result of prefixing '~' to any sentence form is a sentence form;

   iii. the result of inserting '&' or '∨' between any two sentence forms and enclosing the whole in parentheses is a sentence form;

      nothing is a sentence form unless it can be shown to be so on the basis of the above clauses.

Thus '*p*' and '*q*' are sentence forms (clause i) so '~*p*' and '~*q*' are also (clause ii) as are '(*p* & *q*)' and '(*p* ∨ ~*q*)' (clause iii). Then '~~*p*' and '~(*p* ∨ ~*q*)' are also sentence forms (clause ii again), and so on.

It is customary to follow the practice, employed throughout this text, of omitting *outer* parentheses when no ambiguity would result, writing '*p* ∨ *q*' for '(*p* ∨ *q*)' for example, and to replace parentheses with brackets and braces when readability is thereby promoted (for example, '~{[~(*p* & *q*) ∨ *r*] & *s*}' to replace '~((~(*p* & *q*) ∨ *r*) & *s*)'). Of course the parentheses in '~(*p* & *q*)' *cannot* be omitted.

We can now give a technically satisfying definition of a sentence (in our notation): a **sentence** is an expression obtainable from a sentence form by replacing the sentential variables in the latter with sentence letters.

Disjunction   A sentence of the form '*p* v *q*' is true if and only if at least one of its disjuncts is true, and false otherwise.

Negation   A sentence of the form '~*p*' is true if and only if its component is false, and false if its component is true.

We have already noted that

(10) ~(*O* & *B*)

must be distinguished from '~*O* & *B*'. We can notice now, however, that (10) and

(11) ~*O* v ~*B*

may be used interchangeably, and more generally a sentence of the form '~(*p* & *q*)' will always convey precisely the same information as could be conveyed by the corresponding sentence of the form '~*p* v ~*q*'. A conjunction, remember, is true *only* when both components are true and so (10), the negation of a conjunction, will be *false* only when '*O*' and '*B*' are both true. Now (11), a disjunction, is false only when both disjuncts, '~*O*' and '~*B*', are false; hence (11) is false only when their contradictories, '*O*' and '*B*', are *true*. Thus (10) and (11) have the same truth-value regardless of the truth-values of their components: they will both be true so long as '*O*' and '*B*' are not both true, and they will both be false if '*O*' and '*B*' are both true.

A similar relation obtains between

(12) ~*O* & ~*B*

and

(13) ~(*O* v *B*).

Again we have two sentences that have the same truth-value under all conditions, but here both will be true only when '*O*' and '*B*' are both *false*. Since (12) is a conjunction, it will be true only when '~*O*' and '~*B*' are both true, hence only when '*O*' and '*B*' are both false. And since '*O* v *B*' is false only when both disjuncts are, (13)—the negation of '*O* v *B*'—will be true only when '*O*' and '*B*' are both false. More generally, any two sentences of the forms '~*p* & ~*q*' ('not-*p* and not-*q*', in idiomatic English) and '~(*p* ∨ *q*)' ('it is not the case that either *p* or *q*') will be synonymous, and it should be apparent that English contains yet another and more frequently used expression equivalent to both of these: 'neither *p* nor *q*'. Using any of the three locutions, one will have said something true if and only if both '*p*' and '*q*' are false, so sentences of this sort may be symbolized either in the manner of (12) or of (13).

This should give some idea of how conjunction, disjunction, and negation intermesh,[7] but there are other features of our sentential connectives that warrant brief attention as well. First, both '&' and '∨' are *commutative* — that is, the order of the component sentences in conjunctions and disjunctions is immaterial as regards the truth-value of the whole. In this respect both resemble the operation of addition in arithmetic; just as '3 + 2' and '2 + 3' signify the same number, so must '*A* & *B*' and '*B* & *A*' have the same truth-value, as must '*A* ∨ *B*' and '*B* ∨ *A*'. Moreover, conjunction and disjunction are *associative* as well — that is, the placing of parentheses in compounds with three or more components does not affect the truth-value of such compounds. Thus '(*A* & *B*) & *C*' will be true under the same conditions as '*A* & (*B* & *C*)', since each will be true only when all three conjuncts are true. The same considerations hold for '*A* ∨ (*B* ∨ *C*)' and '(*A* ∨ *B*) ∨ *C*', since each of these is false only if all three disjuncts are false. Again there is a parallel with arithmetic, for '(3 + 2) + 6' signifies the same number as '3 + (2 + 6)'.

On the other hand, just as no number is uniquely signified by '6 − 5 + 1', so there is a similar ambiguity with '*A* & *B* ∨ *C*'. Here parentheses are essential, for without them we would have no idea whether this is supposed to be a disjunction whose first disjunct is a conjunction, or a conjunction whose second conjunct is a disjunction. That the difference is important can be seen by letting '*A*' and '*B*' represent false sentences and '*C*' a true one. If the above expression is construed as the disjunction '(*A* & *B*) ∨ *C*', it will be true because its second disjunct is true. Construed as the conjunction '*A* & (*B* ∨ *C*)', however, it will be false owing to the falsity of its first conjunct. These two sentences, therefore, do not have the same truth-value under all conditions, and parentheses are indispensable for distinguishing them. We shall call the connective which determines the basic form of a sentence its **main connective**. Thus, '(*A* & *B*) ∨ *C*' has '∨' as its main connective, while the main connective in '*A* & (*B* ∨ *C*)' is '&'. But '~(*A* & *B*)' is a negation, so its main connective would be '~'.

Ambiguities of these kinds occur less often in natural languages

---

[7]Traditionally, these equivalences are known as **De Morgan's Laws**, and they will be used extensively in Chapter 3.

such as English. From the relative positions of 'either', 'or', and 'and', for example, we can tell that an English sentence of the form 'either p and q or r' is a disjunction with the corresponding sentence 'p and q' as first disjunct. But one of the form 'either p or q and r' would require the same kind of clarification as does 'A v B & C', for one cannot tell from its structure whether it is a disjunction with a conjunction as a component disjunct or a conjunction with a disjunction as a component conjunct. In conversation, such ambiguities are usually alleviated by emphasis or inflection on the part of the speaker. In printed contexts, commas serve the same purpose (we would recognize 'Either Al will go or else Bill will go, and Charlie will go' as a conjunction, for example).

Concerning the '~' we confine ourselves here to a brief remark about *double negation*. When the negation of a sentence is itself negated, the resulting sentence is equivalent to the original sentence alone. Since the negation of a sentence is true if and only if that sentence is false, clearly '~~O' as well as 'O' will be true if and only if '~O' is false. Thus the sign '~' may be applied to a sentence any number of times and any even number of applications will yield a sentence with the same truth-value as the original.

EXERCISES

I.   Translate each of the following English sentences into symbols:
1.   The number four is even and the number five is odd.
2.   The number five is a prime number, and the number four is not.
3.   It is not the case that the numbers four and five are both prime.
4.   The numbers four and five aren't both prime, but the numbers five and seven are.
5.   Neither four nor six is a prime number.
6.   Five is a prime number, but neither four nor six is.
7.   Either six and four are both prime, or else neither is.

II.   Assign truth-values to 'A', 'B', and 'C' such that 'A & ~(B v C)' is true but '(~A & ~B) v (C & B)' is false.

III.   Find a sentence having the same truth-conditions as the conjunction 'A & B' but which contains no occurrence of the connective '&'.

IV.   Which of the following sentences are derivable from the sentence form '~p v (q & ~(p v r))'? For those which are, specify the substitutions for 'p', 'q', and 'r'.
1.   ~B v (A & ~(B v D))
2.   ~B v (A & ~(C v D))
3.   ~A v (~(B & C) & ~(A v D))
4.   ~~A v (B & ~~(A v C))

5. $\sim A \vee (A \mathbin{\&} \sim(A \vee A))$
6. $\sim\sim C \vee (E \mathbin{\&} \sim(C \vee F))$

V. Is the sentence '$\sim\sim A \mathbin{\&} (\sim(B \vee C) \mathbin{\&} \sim A)$' derivable from any of the following? Briefly explain in each case.
1. $p \mathbin{\&} q$
2. $p \mathbin{\&} (\sim q \mathbin{\&} \sim p)$
3. $\sim p \mathbin{\&} (q \mathbin{\&} p)$

Of course, many English words besides 'and' can serve to indicate conjunction and consequently be handled with the ampersand. The word 'but' is so used in

(14) The senator attended the rally but did not make a speech,

for example, since (14) would clearly be true if both component sentences were true and false if either or both were false. Using '*A*' for 'The senator attended the rally', then, and '*S*' for 'The senator made a speech', (14) can be symbolized as the conjunction '$A \mathbin{\&} \sim S$'. Now

(15) The senator attended the rally and did not make a speech

would also be symbolized '$A \mathbin{\&} \sim S$', yet (14) and (15) seem to differ subtly in their meanings — the former would be more appropriate than the latter, for instance, if there were an implicit expectation that the senator *would* perhaps make a speech at the rally. Despite such rhetorical considerations, however, (14) and (15) would be *true* or *false* under precisely the same conditions; as a result, arguments containing (14) will be evaluated no differently from ones just like them except for containing (15) in place of (14). For our purposes, then — putting considerations of nuance and connotation aside — sentences of the form '*p* but *q*' may be rendered in terms of '&'.

Similar remarks apply to a wide assortment of English connectives. Those italicized in the following list, though they may differ in nuance from 'and', all serve to indicate conjunction and may be handled with the ampersand.

Although the weather was bad, the game was not called off.

John, *as well as* Sam, attended the meeting.

Not only will the Mets win the pennant, *but* they will win the Series *as well*.

Johnson was an aggressive president, *whereas* Nixon preferred a low profile.

Despite its controversial status, the bill passed the Senate.

The game was exciting, *albeit* violent.

The notions are simple *yet* the subject is not.

The list is not meant to be exhaustive; indeed, the reader should be able to extend it considerably with a little effort. It should be noted also that conjunction may be indicated not by any particular word or phrase but by punctuation alone: Caesar's "I came, I saw, I conquered" is a three-component conjunction.

And now the task of expressing an exclusive disjunction such as

(16) *Either* we buy groceries *or* we visit your relatives, *but not both*

in terms of the connectives '&', 'v', and '~' is no more than an easy exercise. (16) cannot be put simply as

(16') *G* v *R*,

of course, for the latter is true — but the former false — if both disjuncts are true. So to express (16) fully we rule out this possibility by translating 'but' as '&' and incorporating into (16') the expression '... but not both' from (16):

(16") (*G* v *R*) & ~(*G* & *R*).

Many words besides 'and', then, may be used to indicate conjunction. But we must not jump to the conclusion that such terms always function in this manner. In "None but the strong survive", for example, we do not have a compound sentence at all, 'but' serving merely as a synonym for 'except'.

To further complicate matters, the word 'and' itself can be used nonconjunctively. 'Mary and Jane are sisters' cannot be construed as a conjunction of 'Mary is a sister' and 'Jane is a sister', for these might be true while the first sentence was false. The force of the first sentence is clearly that Mary and Jane are sisters *of one another*. Here, then, 'and' does not serve to connect two sentences; the role it does play in this example will be studied in Chapter 5.

Even where 'and' appears between two sentences as a sentential connective, we need not have a (mere) conjunction. Genuine conjunctions, recall, are commutative — the order of the two conjuncts within the sentence is irrelevant to the truth-value of the whole. But this is not the case with

(17) Jones bought stock and the market soared

or

(18) The market soared and Jones bought stock,

as those who buy at the peak of a bull market can testify. Here 'and' definitely serves to connect two sentences, but plays a temporal role as well. Normally there would be the distinct suggestion with (17) that the event referred to first occurred before that referred to last. In effect 'and' occurs in such sentences in the sense of 'and then', and that sense is not captured by our '&'.

Using '*J*' for 'Jones bought stock' and '*M*' for 'The market soared', we may of course capture *part* of the sense of (17) — and the same part of the sense of (18) as well — with

(17') *J* & *M*,

for if the latter were false — if Jones did not buy stock, or if the market did not soar — (17) and (18) would both be false as well. But it is important to remember that (17') does not tell the *whole* story — only *part* of the force of (17) is reflected in our notation.

When we turn to disjunctions, we find that English provides only trivial alternatives to the 'either ... or ---' construction: the 'either' may be omitted, or the 'or' followed by 'else'. It is worth noting, however, that 'unless' seems sometimes to stand to 'v' as 'and then' stands to '&'. If, for example,

(19) Jones did not buy stock unless his left elbow itched

is true, then

(20) Either Jones did not buy stock or else his left elbow itched

will be true as well. For if the latter were false — if Jones did buy stock but his left elbow did not itch—then (19) is clearly false as well. So at least *part*[8] of the sense of (19) can be expressed (using obvious notation) by

(20') ~J v L.

For many purposes the fact that our symbolic resources are limited is not terribly serious. Thus, we may not be able to translate with full accuracy the premises of the following argument:

Jones bought stock and (then) the market soared.
Jones did not buy stock unless his left elbow itched.
Therefore, Jones' left elbow itched.

We can nonetheless consider a related argument in symbols

J & M
~J v L
Therefore L

The latter is clearly valid—if its premises are both true then its conclusion must be also, for it follows from the first premise that Jones did buy stock and from that and the second that his left elbow itched. And since (17) and (19)—the premises of the English argument—can only be true if the premises are, it is clear that the former must be valid as well.

EXERCISES

I.   Put each of the following sentences into symbols using the suggested notation:
1.   The Hope diamond is valuable, but not priceless. (V, P)
2.   Either the Reds or the Dodgers will win the pennant, but neither will win the Series. (R, D, W, S)
3.   For you I'll climb the tallest mountain, swim the widest ocean, cross the burning desert. (M, O, D)
4.   Despite the fact that it lacked support in Congress, the President decided to introduce the bill and take his chances. (C, P, T)

---

[8]If 'Jones's left elbow itched' is true, it follows that (20) is true also. It is not clear that (19) follows as well, and this is why it is not clear that we can capture the *whole* sense of (19) with (20). We note, however, that when in exercises later on we use the word 'unless', we do so to avoid terminological monotony and the reader may safely take us to mean no more than 'or' by it.

5. Withdrawals will not be honored unless presented by the depositor and accompanied by his passbook. (W, P, A)

6. Either the gold and silver issues will decline or the Dow-Jones index will fall sharply, but in no case will the President's or Congress's decision on the bill be affected. (G, S, D, P, C)

7. The numbers two, four, and six are all even, and at least two of the three are composite. (T, F, S, W, O, I)

8. Although he is willing, the vice-president will not deliver the speech unless it is printed in crayon. (V, D, P)

*II. What can be said about the following argument?
Jones did not bluff on that deal unless he was not dealt a strong hand. Jones was dealt a strong hand on that deal and (then) won the whole pot.
Therefore Jones did not win the whole pot unless he bluffed on that deal.

1.4 Material Conditionals and Biconditionals

At the end of the preceding section we argued that

(20) Either Jones did not buy stock or else his left elbow itched

expresses at least in part the sense of 'Jones did not buy stock unless his left elbow itched' so that

(20') ~J v L

captures at least some of the force of the latter. In many contexts (20) might be paraphrased as

(21) If Jones did buy stock then his left elbow itched.

We shall call (21) a **material conditional** and, following standard practice, refer to 'Jones did buy stock' as its **antecedent** and to 'Jones's left elbow itched' as its **consequent**. Our contention that (20) and (21) stand or fall together can be borne out by arguing informally that (20) and (21) must have the same truth-value. To this end, suppose first that (20) is *true*. Then either Jones did not buy stock or else his left elbow itched. If Jones *did* buy stock, then (20) leaves us with only one alternative: his left elbow itched. Thus, when (20) is true (21) must be also. Now suppose, on the other hand, that (20) is *false*. Then both of its disjuncts, 'Jones did not buy stock' and 'Jones's left elbow itched', must also be false in such a case; anyone who asserts (21) has asserted a falsehood should it turn out that Jones bought stock but his left elbow did *not* itch. So when (20) is false (21) must be false also.

Thus the conditions under which (21) would be true or false coincide with those for (20); and this point may be put by saying that (21) would be false if and only if its antecedent were true and its consequent false, and true otherwise. It is this last feature of (21) that suggests our general

DEFINITION   A **material conditional** whose antecedent is '*p*' and whose consequent is '*q*' is a compound sentence built up from '*p*' and '*q*' (typically, one of the form 'if *p* then *q*'), which is false if and only if '*p*' is true and '*q*' false, and true otherwise.

We shall translate material conditionals from English into symbols by

27

writing '⊃' (called the **horseshoe**) between their antecedents and consequents. (21), for example, will be written

(21') *J* ⊃ *L*.

The term 'material conditional' will be extended to such expressions as (21'), too. As with negations and conjunctions, English provides alternate forms of speech. The 'then' may be replaced with a comma ('*If* Jones did buy stock, his left elbow itched') and the order may be reversed ('Jones's left elbow itched *if* he bought stock'). And often in place of a sentence of the form 'if *p* then *q*' one uses quite different words: 'provided that *p*, *q*', 'in the event that *p*, *q*', or '*p* only if *q*' (for example, '*Provided that* the merchandise is returned, the money will be refunded', '*In the event that* it rains, the game will be postponed', and 'Smith works in the Empire State Building *only if* Smith works in New York City'). It is especially important to notice this last example. Obviously

(22) Smith works in the Empire State Building only if Smith works in New York City

is true. It can only be construed, then, as synonymous with

(22') If Smith works in the Empire State Building then Smith works in New York City,

and it would be a mistake to think that it could be paraphrased by

(23) If Smith works in New York City then Smith works in the Empire State Building;

he might work in the Chrysler Building. (22') and (23), incidentally, are said to be *converses* of one another, and each is called the *converse* of the other. Clearly they need not have the same truth-value.

It follows from our definition of a material conditional, however, that a sentence of the form '*p* ⊃ *q*' does have the feature of being equivalent in truth-value to the corresponding sentence of the form '~*q* ⊃ ~*p*'. We delay establishing this until Section 2.3, where less cumbersome methods of supplying the details will be available, and notice just one example in English where this feature is at work: 'If everybody participated then everybody benefited' and 'If not everybody benefited then not everybody participated' are but two ways of putting one point. A similar relation obtains between 'If Jesse lies and Frank cheats, then the James boys are dishonest' and 'If Jesse lies, then if Frank cheats then the James boys are dishonest'. Again, the details involved in showing generally that a sentence of the form '$(p \ \& \ q) \supset r$' must have the same truth-value as the associated sentence of the form '$p \supset (q \supset r)$' are postponed until Section 2.3.

We emphasize that our concern is with *material* conditionals, with sentences of the form 'if p then q' in which 'if ... then ---' is so used that their truth conditions coincide with those of the corresponding sentences of the form 'either not-p or q'. It would be nicer, of course, if *material* conditionals could simply be identified with conditionals generally, that is, with 'if ... then ---' sentences (and their idiomatic variants). But there are many sentences containing 'if ... then ---' whose truth or falsity seems to depend on more than the truth-values of their components. To take but one example (see Section 2.3 for more), the material conditional

This piece of wire is made of copper $\supset$ this piece of wire will not conduct electricity

must be counted as true so long as its antecedent, 'This piece of wire is made of copper', is false. So it will be true if the piece of wire in question is made, for example, of steel. It seems plausible to suppose, however, that the English sentence "If this piece of wire is made of copper then this piece of wire will not conduct electricity" is false regardless of the actual composition of the wire.

Material conditionals, then, are but one sort of conditional one might study. We single them out for study in this book partly because they yield most readily to elementary methods, and partly because many of their features are shared by most conditionals.

One other sentential connective occurs often enough in arguments to justify introduction of a special symbol: '... if and only if ---' ('... just in case ---' is a synonym). Of course such sentences as

(24) Art goes if and only if Betsy goes

and

(25) Art goes just in case Betsy goes

can be handled with the connectives already introduced — we have only to notice that each may be paraphrased by

(26)   If Art goes then Betsy goes, and if
Betsy goes then Art goes

and so can be rendered as

(26′) $(A \supset B)$ & $(B \supset A)$.

Alternately, we could notice that (24)–(26) may all be paraphrased by

(27)   Either Art goes and Betsy goes, or else
Art doesn't go and Betsy doesn't go,

and so render any of them as

(27′) $(A \ \& \ B) \lor (\sim A \ \& \ \sim B)$.

Instead we will introduce the symbol '$\equiv$' (the **triple bar**), using it in effect as an abbreviation for the expression '... if and only if ---' as it occurs in (24) so that the latter may be written simply

(24′) $A \equiv B$.

We shall call (24) and (24′) **material biconditionals**, construing them as true when their components have the same truth-value and false when their components differ in truth-value.[9]

EXERCISES

I. Put each of the following sentences into symbols, using the suggested notation:
1. If the sun's rays hit the polar regions squarely, then the polar regions will turn warm. (R, W)

---

[9]To extend the notion of a sentence form, we now add to the definition in footnote 6 a new clause:
   iv. the result of inserting '$\supset$' or '$\equiv$' between any two sentence forms and enclosing the whole in parentheses is a sentence form.

2. If Marr misses his tap-in and the spectators groan loudly, Palmer's concentration will be disturbed. (M, G, D)
3. If the spectators let out a spontaneous shout if Marr misses his tap-in, Palmer will win. (S, M, P)
4. If his heart stops beating, the patient will live if the doctors apply electric shock. (S, L, E)
5. Withdrawals made after the first ten days of an interest-paying month will be honored only if the depositor has given the bank written notice of his intent to withdraw as well as the amount of the withdrawal. (W, I, A)
6. If time is not finite and not circular then time has neither a beginning nor an end. (F, C, B, E)
7. Jones will graduate if he pays his tuition and passes the requisite number of courses, but he will graduate only if he pays his tuition. (G, T, C)
8. Jones will graduate whether or not he pays his tuition. (as above)
9. I'll do the job if, but only if, nobody else will. (I, N)
10. If the job has to be done, then I'll do it for you just in case you don't find anyone else to do it. (J, I, F)

II. Show that '⊃' is not commutative by finding sentences 'p' and 'q' such that '$p \supset q$' and '$q \supset p$' have different truth-values.

III. Show that '⊃' is not associative by finding sentences 'p', 'q', and 'r' such that '$p \supset (q \supset r)$' and '$(p \supset q) \supset r$' have different truth-values.

IV. Is '≡' either commutative or associative?

V. Are there sentences 'p' and 'q' such that '$p \equiv \sim q$' and '$\sim(p \equiv q)$' have different truth-values? How about '$p \equiv \sim q$' and 'either p or q, but not both'? Show your work.

1.5 Scope of Our Inquiry

In Sections 1.2 and 1.3 we examined *sentences*, focusing on sentential connectives like 'and' and 'or'. Typically, the truth-values of compounds containing them are determined solely by the truth-values of their components. It was just this characteristic we relied on, in fact, in defining such special connectives as '&' and 'v'.

Compound sentences, whether in English or symbols, whose truth-values depend entirely on the truth-values of the sentences occurring in them—are called **truth-functional**. And it is important to realize that English and other natural languages also contain **non-truth-functional** sentences, compounds whose truth-values are *not* dependent solely upon the truth-values of their components. The connectives used in forming such compound sentences function quite differently, then, from those we have concentrated on in the last two sections; indeed, explaining just how they work is in many cases still a topic for philosophical investigation. 'And then'[10] provides one example of such a connective, and 'because' another. The sentence

The chairman resigned *because* he was in ill health,

for example, is not truth-functional. Even if 'The chairman resigned' and 'The chairman was in ill health' are both true, the truth-value of the compound itself is not determined—it may be true or it may be false, depending on whether ill health was in fact the *reason* for the resignation. This is quite unlike what happens with truth-functional sentences; if the truth-values of 'p' and 'q' are given, the truth-values of the corresponding sentences of the forms 'p & q', 'p v q', and so on, are thereby determined.[11]

Verbs of *psychological attitude*—such as 'believes', 'doubts', 'hopes'—afford further examples of non-truth-functional compounds. For the truth-value of the sentence

Chris believes that the world is egg-shaped,

---

[10]The reader interested in acquiring an understanding of how investigations of such connectives proceed might examine G. E. M. Anscombe, "Before and After," *Philosophical Review*, vol. 73 (1964), pp. 3–24, and G. H. von Wright, "And Then," *Comm. Phys. Math. of the Finnish Society of Sciences*, vol. 32, no. 7 (1966).

[11]To count as truth-functional, a compound sentence must be such that its truth-value is determined by *every* combination of truth-values for its components. Thus even though the truth-value of this sentence *is* determined when 'The chairman resigned' is true and 'The chairman was in ill health' is false, the fact that it is not determined when both are true suffices to make it non-truth-functional.

(a compound built up from 'The world is egg-shaped' and 'Chris believes that …') is in no way determined by the truth-value of the component sentence. That someone does or does not hold a certain belief can hardly be established by merely noting the truth-value of a sentence expressing the belief.

While the study of non-truth-functional compounds would be highly challenging, it presupposes a good deal of training in the more elementary areas of logic and so lies beyond the scope of an introductory text. We mention such sentences here only for contrast with our present concern: truth-functional compounds. Arguments containing the latter form the simplest and most basic subject matter within our ken, and it is with them that we must begin.[12]

It remains only to say a few words about the devices ordinarily used in English for marking off the conclusion of an argument. Often this is done by prefixing the word 'therefore' to the conclusion, as in

If the set of integers is finite, it has a largest member.
The set of integers does not have a largest member.
Therefore the set of integers is not finite.

Many other terms can do the job as well; 'consequently' in the place of 'therefore' would serve the same purpose, as would 'thus' or 'hence' or 'so'. Such terms do not function here as sentential connectives, but serve rather to suggest that a certain relation obtains *between* sentences. Typically, these terms are used to indicate that the writer or speaker is attempting to establish the truth of one claim on the basis of the truth of certain others — more specifically, that the former follows from the latter. We shall call them **conclusion indicators**.

It is also possible to express an argument in a single (compound) sentence, the *premises* being marked off explicitly by a **premise indicator** such as 'for':

The set of integers is not finite; for if it is finite it has a largest member, but the set of integers does not have a largest member.

In such cases one or more premises are often left unstated, having been assumed to be sufficiently obvious for the reader to supply on his own:

The set of integers is not finite, for it does not have a largest member.

---

[12]Since non-truth-functional compounds are beyond our purview, no special symbols will be introduced for words like 'and then', 'because', and 'believes'. Instead, compounds involving such terms must be represented by single sentence letters, as if they were simple sentences.

Again other terms will do the job, 'for' being replaceable with similar effect here by such words as 'since', 'inasmuch as', and—an old friend, used in a new sense—'because'.[13]

It will be helpful, in what follows, to have one standard way of marking off conclusions. We shall use the standard symbol '∴' for this purpose.

If the set of integers is finite, it has a largest member.
The set of integers does not have a largest member.
∴ The set of integers is not finite.

Custom suggests that '∴' be read 'therefore'.

EXERCISES

I.   Some of the following sentences are truth-functional and some are not. Express in symbols those which are (those which are not must be represented by single sentence letters, e.g., 'A').

1.   Even though the roads were icy, Sam arrived on time.
2.   He was her man, but he done her wrong.
3.   While no one complained, they weren't very happy either.
4.   Jim thinks a sea dragon inhabits the Wabash.
5.   Swarthmore will not defeat Penn unless God intervenes.
6.   Swarthmore will not defeat Penn even if God intervenes.
7.   Should the game be called off, both the players and the fans will be disappointed,
8.   Though many have tried none have succeeded, but I feel sure someone will, and soon.
9.   Although Little Z and Big Yama were caught in the vice raid, neither will be prosecuted unless the D.A. is incorruptible.
10.   In the event of rain, the game will be postponed.
11.   It is possible but not likely that the Cubs will win.
12.   Scott will come as long as he doesn't have to wear a tie.
13.   Since no one objected, the bill passed without modifications.
14.   Sam and Irma have been man and wife for three years.
15.   God knows that $2 + 2 = 4$.
16.   The owl and the pussy cat went to sea together.
17.   If Swarthmore wins, I'll eat my hat.

---

[13]Presumably it goes without saying that these terms do not *always* serve to distinguish premises from conclusion, nor to indicate that an argument is being given. In "Tom was so late he missed the train" and "I have not seen him since nine o'clock" no arguments are involved and neither 'since' nor 'so' functions as a premise indicator.

18. It's necessarily true that $12 - 7 = 5$.
19. The sentence 'Chicago is in Illinois' is true.
20. 'Twas the night before Christmas, and all through the house, not a creature was stirring, not even a mouse.

II.   Construct a list of additional phrases which may serve to mark off the premises or conclusion of an argument, and illustrate this use for each.

CHAPTER TWO

SENTENTIAL LOGIC
Semantic Methods

2.1 Truth-Tables

In Chapter 1 we noticed that a sentence of the form '$\sim p$' is true if and only if the negated sentence is false; a sentence of the form '$p \& q$' is true if and only if both conjuncts are true; and a sentence of the form '$p \vee q$' is true if and only if it has at least one true disjunct. Each type of compound was thus explained in terms of the conditions under which it would be true, and these conditions may be set out more explicitly in tabular form. The following **truth-tables** lay out the truth-conditions for negations, conjunctions, and disjunctions respectively and therefore may formally define the connectives '$\sim$', '$\&$', and '$\vee$' of our symbolism:

(I)	p	$\sim p$
	T	F
	F	T

(II)	p	q	$p \& q$
	T	T	T
	T	F	F
	F	T	F
	F	F	F

(III)	p	q	$p \vee q$
	T	T	T
	T	F	T
	F	T	T
	F	F	F

Table (I) has but two rows, since there are just two truth-values the component of a negation may have: it may be true or it may be false. Conjunctions and disjunctions typically have two components, however, so their truth-tables require four rows: one for each of the possibilities (1) that both components are true, (2) that the first is true but the second false, (3) that the first is false but the second true, and (4) that both are false. As table (II) shows, conjunctions are true only in the first of these cases; but a disjunction is shown by table (III) to be true on each of the first three conditions — it has a **T** in all but the last row of its column where both disjuncts are assigned **F**. Each row of these truth-tables, then, gives: first, an assignment of truth-values to the components of a truth-functional compound; and second, the truth-value of the compound itself *for* that assignment.

Now for any truth-functional compound sentence containing one or more of '&', 'v', and '~', we can construct a truth-table exhibiting its truth-conditions by using the above tables as our guide. Consider, for example, the sentence '~A v (B & A)'; from table (I) we can ascertain for each row the truth-values for the component '~A', and table (II) allows us to do the same for the conjunction 'B & A'. Thus having the truth-value assignments in each of the rows for both disjuncts, the assignments for the entire disjunction are obtained using table (III).

A	B	~A	B & A	~A v (B & A)
T	T	F	T	T
T	F	F	F	F
F	T	T	F	T
F	F	T	F	T

A compound with three components will require a truth-table of eight rows to set out the conditions under which it will be true and false, since there are exactly eight possible assignments of truth-values to any triad of sentences. Similarly, a four-component compound requires 16 rows, a five-component compound 32 rows, and so on; the number of rows doubles as the number of components is increased by one.[1]

Let us consider more closely compounds with three components; what applies to them applies by extension to those with four or more. In particular, we will examine two such compounds cited back in Chapter 1: 'A & (B v C)' and '(A & B) v C'. There is an easy technique for making sure that the table contains all eight truth-value combinations — that none have been duplicated or omitted. Assign all **T**'s to the first four rows beneath the first simple component ('A' in this case) and all **F**'s to the last four. Then alternate two rows of **T**'s and two of **F**'s beneath the second component, and finally alternate **T**'s and **F**'s beneath the last. To determine what truth-values 'A & (B v C)' will have for each of these eight cases, we first obtain the assignments for the right-hand conjunct 'B v C', putting a **T** in every row of its column where either or both of

---

[1] In general, then, a sentence with n components (n a positive integer) requires a truth-table with 2^n rows. Note that whereas '(A & B) v C' has three components and so requires an eight-row table, '(A & B) v A' has but *two* components (though 'A' has more than one *occurrence*) and requires only a four-row table.

'*B*' and '*C*' are assigned **T**, and an **F** in the others. We then determine the truth-values for the entire conjunction by placing a **T** in every row of its column where both '*A*' and '*B* v *C*' are assigned **T**, and an **F** in the remaining rows. By the same token, the truth-values for '(*A* & *B*) v *C*' are obtained by first getting the truth-values for the component conjunction '*A* & *B*' and then putting a **T** in just those rows where a **T** has been assigned to at least one of the disjuncts, '*A* & *B*' and '*C*':

A	B	C	B v C	A & (B v C)	A & B	(A & B) v C
T	T	T	T	T	T	T
T	T	F	T	T	T	T
T	F	T	T	T	F	T
T	F	F	F	F	F	F
F	T	T	T	F	F	T
F	T	F	T	F	F	F
F	F	T	T	F	F	T
F	F	F	F	F	F	F

Comparing the columns for '*A* & (*B* v *C*)' and '(*A* & *B*) v *C*', we see that they are assigned different truth-values in rows 5 and 7. Thus, this table shows graphically what was noted in Chapter 1, that these two sentences convey quite different information. When, for example, the sentences represented by '*A*' and '*B*' are false and that by '*C*' true, '(*A* & *B*) v *C*' will be true but '*A* & (*B* v *C*)' false. Thus these sentences are not true under *all* the same conditions.

Let us now construct truth-tables for '*A* v ~*A*' and '~*A* v (*B* v *A*)':

A	~A	A v ~A
T	F	T
F	T	T

A	B	~A	B v A	~A v (B v A)
T	T	F	T	T
T	F	F	T	T
F	T	T	T	T
F	F	T	F	T

The truth-table columns for both sentences contain nothing but **T**'s, and they are thus true under *all* conditions — no circumstance whatever will render them false. Any truth-functional sentence that is true no matter what truth-values we may suppose its simple component sentences to have will be called a **tautology**.

DEFINITION   A truth-functional sentence is a **tautology** if and only if it takes the truth-value **T** for every assignment of truth-values to its component sentence letters.

We shall use the sign '$\models$' in abbreviating the assertion that a given sentence is a tautology.[2] Thus, instead of writing

The sentence '$A \lor \sim A$' is a tautology

we shall simply write

$\models A \lor \sim A$.

Tautologies obviously form a unique subclass of true sentences, for they are true not because the facts are one way rather than another, but because of their *form* — in the case of '$A \lor \sim A$', because it is of the form '$p \lor \sim p$', that is, a disjunction one of whose disjuncts is the negation of the other, thus ensuring that there will always be a true disjunct. Since they are true independently of how the world happens to be constituted, tautologies are often said to be *necessarily* true, or in philosopher's jargon, true in "all possible worlds."

The tautology '$A \lor \sim A$' is true on purely formal grounds. And it should be obvious that no matter what sentence is substituted for 'p' in the associated form '$p \lor \sim p$' the result must be a truth. Indeed, we saw that it is *because* '$A \lor \sim A$' is of the form '$p \lor \sim p$' that it is true on formal grounds alone, and the same clearly holds for the tautologies '$B \lor \sim B$' and (making complex substitutions for 'p') '$\sim C \lor \sim\sim C$' and '$(A \& B) \lor \sim(A \& B)$'. We shall call any sentence form that yields nothing but tautologies upon the substitution of sentences (simple or compound) for its variables a **tautologous form** (the same sentence must of course replace the same variable throughout). To be a tautology, then, a sentence must be derivable from at least one tautologous form, but it must not be thought that tautologies are derivable *only* from tautologous forms. For while '$A \lor \sim A$' is derivable from the tautologous form '$p \lor \sim p$', it is also derivable from the nontautologous '$p \lor q$' (substituting 'A' for 'p' and

---

[2] The sign '$\models$' should *not* be confused with the sentential connectives that are here our prime objects of study. We use it only to save ourselves verbiage. Readers should compare '$\models$' with the sign '$\vdash$' of Section 3.1. Note incidentally that '$\models \sim A \lor (B \lor A)$' is true, but '$\models \sim A \lor (B \& A)$' is not.

'~*A*' for '*q*'). Whereas tautologous forms yield only tautologies upon substitution, nontautologous forms can yield both tautologies and nontautologies.

Since their truth is independent of the facts, tautologies offer us no real information as to how things are (or are not). In uttering "It is now raining," one offers some information (true or false) about the environment, but in "Either it is now raining or it is not," no information is given concerning the present state of the weather. All tautologies are "trivial" in this sense owing to their lack of factual content, and the reader may be hard put to imagine an actual context where it would not be pointless to utter one. Trivial though they seem, however, tautologies nonetheless are an important tool in the logician's repertoire. Their necessary truth gives them a theoretical significance which will be examined later on.

Now let us consider the truth-table for the negation of '*A* v ~*A*':

A	~*A*	*A* v ~*A*	~(*A* v ~*A*)
T	F	T	F
F	T	T	F

The sentence '~(*A* v ~*A*)' has nothing but **F**'s in its column and shall be called a **contradiction** — there is no condition under which it could possibly be true. Clearly the negation of any tautology is a contradiction (and vice versa), and just as a tautology is necessarily true so a contradiction is necessarily false — its falsehood being due entirely to its form. Thus sentence forms like '~(*p* v ~*p*)' and '*p* & ~*p*' which yield only contradictions upon substitution shall be termed **contradictory forms**. Everything that has been said above concerning tautologies and tautologous forms applies *mutatis mutandis* to contradictions and contradictory forms.

It was noted in the preceding chapter that the sentences '~*O* & ~*B*' and '~(*O* v *B*)' convey the same information, which may be expressed in English by 'God is neither omnipotent nor benevolent'. A truth-table with four rows will establish this in a more lucid fashion than the argument given in Chapter 1. The table shows the two sentences to have exactly the same arrangement of **T**'s and **F**'s in their respective columns — thus showing them to be true under just the same conditions:

O	*B*	~*O*	~*B*	~*O* & ~*B*	*O* v *B*	~(*O* v *B*)
T	T	F	F	F	T	F
T	F	F	T	F	T	F
F	T	T	F	F	T	F
F	F	T	T	T	F	T

Sentences true under the same conditions have thus far been characterized as *saying the same thing* or *conveying the same information*. We shall now introduce a special term to replace these rather vague phrases.

DEFINITION   Two truth-functional sentences are **equivalent** if and only if, for each assignment of truth-values to their component sentence letters, both receive the same truth-value.

It follows from this definition that all tautologies are equivalent to one another and to no nontautologies; and similarly that each contradiction is equivalent to all other contradictions but to no noncontradictions.

Moreover, just as it is in virtue of their forms that certain sentences are tautologies, so is it in virtue of their respective forms that certain pairs of sentences are equivalent. Any sentence whatever of the form '$\sim p$ & $\sim q$' will be equivalent to the corresponding sentence of the form '$\sim(p \lor q)$'. For the truth-table in effect shows us that the negation of *any* disjunction will have the same truth-conditions as the corresponding conjunction of the negated disjuncts; the particular sentential content makes no difference, since any sentences whatever could have been used in place of '*O*' and '*B*' in constructing the truth-table. Thus, we shall extend our use of 'equivalent' and say that the sentence forms '$\sim p$ & $\sim q$' and '$\sim(p \lor q)$' are equivalent as well. Plainly, then, when two *sentences* are equivalent, they are derivable from a pair of equivalent sentence forms through the substitution of their component sentences for the variables—the same sentence replacing the same variable throughout.[3] The foregoing also shows that the use of truth-tables is, from a practical standpoint, unaffected by the type of expression—whether sentential variable or sentence letter—employed in its construction. The following table, for example, differs little in appearance from the preceding one, but it shows the equivalence of the forms '$\sim p$ & $\sim q$' and '$\sim(p \lor q)$':

p	q	$\sim p$	$\sim q$	$\sim p$ & $\sim q$	$p \lor q$	$\sim(p \lor q)$
T	T	F	F	F	T	F
T	F	F	T	F	T	F
F	T	T	F	F	T	F
F	F	T	T	T	F	T

---

[3]Two important characteristics of the *equivalence* relation are:
1. Every sentence (sentence form) is equivalent to itself.
2. If one sentence (sentence form) is equivalent to a second and the second to a third, then the first is equivalent to the third.

EXERCISES

I. Determine whether the following are tautologies, contradictions, or neither:
1. A v ($\sim A$ & A)
2. $\sim(A$ & $B)$ v A
3. $(A$ & $\sim B)$ v $\sim[\sim(B$ & $C)$ & $(C$ & $A)]$
4. $\sim[(A$ v $B)$ & $C]$ v $\sim(D$ & $B)$
5. $(A$ & $\sim B)$ & $[\sim A$ v $(B$ & $A)]$

II. Using truth-tables, show that the following pairs of sentence forms are equivalent:
1. p & q; q & p
2. p v q; q v p
3. p & $(q$ & $r)$; $(p$ & $q)$ & r
4. p v $(q$ v $r)$; $(p$ v $q)$ v r
5. p & $(q$ v $r)$; $(p$ & $q)$ v $(p$ & $r)$
6. p v $(q$ & $r)$; $(p$ v $q)$ & $(p$ v $r)$
7. p; $\sim\sim p$
8. $\sim p$ v $\sim q$; $\sim(p$ & $q)$

III. 1. "If a conjunction is a tautology, then each of its conjuncts must also be a tautology." True or false? Explain.
2. "If the *negation* of a conjunction is a tautology and the *negation* of one of its conjuncts is a contradiction, then the other conjunct must be a contradiction." True or false? Explain.

2.2 Truth-Tables for '⊃' and '≡'

In the preceding chapter English sentences containing 'if ... then ---' and its idiomatic variants were introduced and *material* conditionals[4] singled out for special study. It will be recalled that the latter are expressed in symbols by the sentential connective '⊃', and are false only when their antecedents are true but their consequents false. The following truth-table serves to spell out the conditions under which such sentences are counted as true or false, and so may be regarded as defining '⊃' (note also the equivalence of '$p \supset q$' with '$\sim p \vee q$'):

p	q	$p \supset q$	$\sim p$	$\sim p \vee q$
T	T	T	F	T
T	F	F	F	F
F	T	T	T	T
F	F	T	T	T

Because of the similarity noted earlier between sentences of the form 'if p then q' and those of the form 'either not-p or q', those of the first sort may, with certain reservations to be mentioned presently, be expressed in symbols as material conditionals of the form '$p \supset q$'. But it should not be supposed, nor is it necessary for our purposes to suppose, that the full meaning of an English conditional is captured in our notation. The variables in '$p \supset q$' may be replaced with any two sentences whatever, no matter how relevant or irrelevant their particular contents might be to one another. Thus where 'R' represents 'The world is round' and 'E' represents '$2 + 2 = 4$', both '$R \supset E$' and '$\sim R \supset \sim E$' are considered true since in the former case both antecedent and consequent are true and in the latter both are false. However, the associated English conditionals,

(1) If the world is round then $2 + 2 = 4$

(2) If the world is not round then $2 + 2 \neq 4$,

sound very odd indeed, and there would seem to be no more reason for regarding them as true than as false. Moreover, we ordinarily would have no reason for ever using (1) or (2) unless perhaps in the case of (2) we wish to state with great emphasis the truism that the world is round. The strangeness of conditionals like (1) and (2) stems partly from the fact that

---

[4]This term, and some of the ensuing discussion, is borrowed from W. V. Quine. See his *Methods of Logic* (ed. 3), New York: Holt, Rinehart and Winston, 1972.

the truth-values of their component sentences are beyond question whereas normally one who utters a conditional is uncertain of the truth-values of its antecedent and consequent. But more importantly, it stems from the fact that there is no relevant "connection" between antecedent and consequent, whereas normally we do not utter a conditional unless we believe there to be such a connection. In these respects then, (1) and (2) differ markedly from

(3)   If the Governor gave the speech, the audience was bored.

Here, though we know about the Governor's lack of oratorical skill, we are nonetheless unsure of who spoke and what the mood of the audience was. And generally, when we assert something of the form 'if p then q', though we may not know the truth-values of 'p' and of 'q', we have some reason for *denying* the conjunction 'p and not-q' — in the present case, our knowledge of the Governor's speaking deficiencies would lead us to deny that he spoke and yet that the audience was *not* bored. Herein we can see why '$p \supset q$' and '$\sim(p \& \sim q)$' are to be construed as equivalent.

So the material conditional form differs from many ordinary English conditional sentences like (3) in that it carries no presumption of relevance of content between antecedent and consequent. No underlying principle or other knowledge connecting the subject matter of the two components need be presumed. Conditionals like (1) and (2) are thus assigned truth-values in accordance with the truth-table for '⊃'.

Unnatural though this may seem, it is often harmless and of no great consequence. What is important is that in going from 'if p then q' to the material conditional '$p \supset q$', we abstract from whatever "connection" is felt to obtain between antecedent and consequent. We concentrate solely on the distinctive truth-functional dimension of the English sentence: that it denies that conjunction consisting of the material conditional's antecedent and the negation of its consequent. No claim of synonymy, then, is made in going from 'if p then q' to '$p \supset q$'. For our purpose of analyzing arguments in which conditionals like (3) may occur, no meaning need be ascribed to 'if ... then ---' beyond that specified in the truth-table.

Now there are many types of English conditionals, and we may distinguish them according to the sort of relevant connection obtaining between antecedent and consequent. Often, this connection is best construed in terms of one or another kind of *necessity*. The following conditionals typify three varieties of necessary connection:

(4)   If all men are mortal and Socrates is a man, then Socrates is mortal

(5)   If John is a bachelor, then he is unmarried

(6)   If this match is heated to 300°F in the presence of oxygen, then it will
      ignite.

> (4) might be termed a *logical* conditional, for the necessary connection
> alleged to hold between antecedent and consequent is one of premise to
> conclusion in a valid argument. In (5), the connection between antecedent
> and consequent is one of definition — 'bachelor' is defined as 'unmarried
> man'. (6) might be termed a *causal* conditional in that the necessity which
> connects antecedent and consequent is one of cause to effect — being in
> the presence of oxygen at a certain temperature is a cause of combustion.
> These varieties of conditional may be spelled out more explicitly as
> follows:

If p then (necessarily it follows that) q

If p then (necessarily by definition) q

If p then (necessarily as cause to effect) q

> With each of these, the consequent not merely *is* true given the truth
> of the antecedent, but *must* be; and 'must' here conveys either logical,
> definitional, or causal necessity.[5] It is from such kinds of necessary
> connection that we abstract when translating such conditionals as '$p \supset q$',
> and it is comforting to know that despite the abstraction, there are count-
> less arguments containing conditionals such as (4), (5), and (6) whose
> validity or invalidity may be determined solely by truth-functional
> techniques. A simple and obvious example would be:

If this wire is made of copper, it will conduct electricity.
This wire is made of copper.
Therefore this wire will conduct electricity.

> Although the first premise is a causal conditional, and hence asserts far
> more than the material variety of our symbolism, when we translate this
> argument by replacing 'if ... then ---' with '$\supset$' its validity is easily estab-
> lished by the techniques to be presented in the next section of this chapter.

---

[5]While a study of logical and causal conditionals is a highly interesting philosophical enter-
prise, it is unfortunately beyond the scope of an introductory text. Readers interested in the
analysis of logical conditionals are advised to consult C. I. Lewis and C. H. Langford,
Symbolic Logic, New York: Dover, 1932, and also A. Anderson and N. Belnap, "The Pure
Calculus of Entailment," *Journal of Symbolic Logic*, vol. 27 (1962), pp. 19–52. Those interested in causal conditionals should look into A. Burks, "The Logic of Causal Propositions,"
Mind, vol. 60 (1951), pp. 363–382. The reader will find, incidentally, that these works
presuppose a thorough familiarity with the material conditional.

Moreover we saw earlier that 'if ... then ---' and '⊃' share some important features. For example, it was noted in Section 1.4 that English conditionals must be distinguished from their converses; (3), for example, may be true while its converse,

(7)   If the audience was bored, the Governor gave the speech,

is false — their boredom being the result of another equally ineffective orator. The following truth-table shows that '⊃' behaves in exactly the same fashion, and also fulfills Chapter 1's promise to establish the equivalence of '$p \supset q$' and '$\sim q \supset \sim p$':

p	q	$p \supset q$	$q \supset p$	$\sim p$	$\sim q$	$\sim q \supset \sim p$
T	T	T	T	F	F	T
T	F	F	T	F	T	F
F	T	T	F	T	F	T
F	F	T	T	T	T	T

The equivalence of '$(p \,\&\, q) \supset r$' and '$p \supset (q \supset r)$' is left as an exercise.

The *material biconditional* was also introduced in the first chapter, and the following truth-table formally defines the connective '≡' as well as demonstrating the equivalence of '$p \equiv q$' and '$(p \supset q) \,\&\, (q \supset p)$':

p	q	$p \equiv q$	$p \supset q$	$q \supset p$	$(p \supset q) \,\&\, (q \supset p)$
T	T	T	T	T	T
T	F	F	F	T	F
F	T	F	T	F	F
F	F	T	T	T	T

Thus, as noticed in Section 1.4, a sentence of the form '$p \equiv q$' is true if and only if both components have the same truth-value. The problems noted earlier in translating English conditionals apply *mutatis mutandis* to the translation of English biconditionals into material ones. No claim is made that '$p \equiv q$' captures the full meaning of its English counterpart, but only that it isolates a distinctive truth-functional aspect so that, in general, validity is preserved in going from arguments with 'if and only if' to '≡'. Moreover, the important properties of commutativity and associativity are shared by 'if and only if' and '≡'; '$p \equiv q$' is equivalent to '$q \equiv p$', and '$p \equiv (q \equiv r)$' is equivalent to '$(p \equiv q) \equiv r$'.

EXERCISES

I. Using truth-tables, show the following pairs equivalent:
    1. $(p \mathbin{\&} q) \supset r$; $p \supset (q \supset r)$
    2. $p \equiv q$; $(p \mathbin{\&} q) \vee (\sim p \mathbin{\&} \sim q)$

II. Is '$p \equiv \sim q$' equivalent to '$\sim(p \equiv q)$'? To the negation of 'either p or q but not both'? Show your work.

III. To keep in practice, put each of the following into symbols using the suggested notation:
    1. If either Carty or Oliva wins the batting crown, both the press and the fans will be happy. (C, O, P, F)
    2. Sam will come if Charlie does, but he won't stay long unless Emma is there. (S, C, L, E)
    3. If either Bench or Rose does not perform up to expectations, then the Reds will win the pennant only if the pitching improves. (B, R, P, I)
    4. I will not run if nominated, I will not serve if elected. (R, N, S, E)
    5. Cohen will be the next Supreme Court Justice if and only if the President recommends him and the Senate approves, but provided that the Old Guard can muster the votes, the Senate will not approve. (C, P, S, O)

2.3 Testing for Validity: The Truth-Table Method

We are now in a position to test truth-functional arguments for validity through the use of truth-tables. (By a **truth-functional argument** we mean one whose premises and conclusion are either simple sentences or else truth-functional compound sentences.) We begin with an elementary example:

Either the stock market will not improve or else the government will both lower interest rates and increase spending. However, the government either will not lower interest rates or else it will not increase spending. Therefore the stock market will not improve.

We shall first put the argument into symbols, using 'S' for 'The stock market will improve', 'L' for 'The government will lower interest rates', and 'I' for 'The government will increase spending'.

(I) $\sim S \lor (L \& I)$
$\sim L \lor \sim I$
$\therefore \sim S$

We know that valid arguments always carry us from true premises to true conclusions so that it is impossible for the former all to be true when the latter is false. Moreover, we know that the truth-values of the premises '$\sim S \lor (L \& I)$' and '$\sim L \lor \sim I$', as well as the conclusion '$\sim S$', are determined once we assign truth-values to the letters 'S', 'L', and 'I'. So if we run through *all* possible assignments of truth-values to these letters and there is *no* combination resulting in true premises and false conclusion, the argument must be valid. That is, the argument's validity emerges from the fact that no matter what truth-values we may suppose 'S', 'L', and 'I' to have, there is no case in which all premises are rendered true but the conclusion false. We know also that a truth-table of eight rows will give us all the required truth-value combinations for three sentence letters, each row giving us a particular assignment of truth-values. So if we construct columns for the premises and conclusion, the validity of the argument is demonstrated by there being no row of the table wherein both premises are assigned **T** and the conclusion **F**.

Comparing the columns for '$\sim S \lor (L \& I)$', '$\sim L \lor \sim I$', and '$\sim S$' in the truth-table on the following page, we can find no row in which the first two are assigned **T** and the third **F**. Notice that it is only in the last three rows that both premises are in fact assigned **T**, but since the conclusion also has a **T** in these rows, the argument is valid.

S	L	I	~S	L & I	~S v (L & I)	~L	~I	~L v ~I
T	T	T	F	T	T	F	F	F
T	T	F	F	F	F	F	T	T
T	F	T	F	F	F	T	F	T
T	F	F	F	F	F	T	T	T
F	T	T	T	T	T	F	F	F
F	T	F	T	F	T	F	T	T
F	F	T	T	F	T	T	F	T
F	F	F	T	F	T	T	T	T

Of course an invalid argument can carry us from true premises to a false conclusion, and this means there is at least one truth-value assignment to the component letters making all premises true and the conclusion false. Therefore, in the truth-table for an invalid argument, *at least one* row will have all **T**'s assigned to the premises and an **F** to the conclusion. As an example, let us consider the argument in which premise two of argument (I) becomes the conclusion and vice versa:

(II)   ~S v (L & I)
    ~S
    ∴ ~L v ~I

The invalidity of (II) is shown by the fifth row of the above table. There, both '~S v (L & I)' and '~S' are assigned **T** but '~L v ~I' is assigned **F**. All that is required is that there be at least one such row, for invalid arguments are those in which a true conclusion does not *always* accompany true premises.

We shall apply the term **truth-table valid** to any argument whose conclusion is never assigned an **F** in a row where all its premises are assigned **T**'s, and it will be convenient to use the sign '⊨' to express this notion.[6] Thus instead of writing

The argument with premises '~S v (L & I)' and '~L v ~I' and conclusion '~S' is truth-table valid

---

[6]The reader may wish to compare this definition and the sign '⊨' with the definition of *deductive validity* and '⊢' in Section 3.1. He should also compare its use here with that of Section 2.1. There it was used to say of a *sentence* that it is a tautology (that is, no row of its truth-table contains an **F**). Here it is used to say of an *argument* that it is valid (no row of *its* truth-table contains an **F** for its conclusion when all premises are assigned **T**'s).

we shall simply write

~S v (L & I), ~L v ~I $\models$ ~S.

The sign '$\models$' is not a sentential connective but merely a shorthand device for expressing the relation between premise(s) and conclusion in a truth-table valid argument. We may now define this notion as follows:

DEFINITION   A truth-functional argument is **truth-table valid** if and only if there is no row of its truth-table in which all its premises are assigned the truth-value **T** but its conclusion the truth-value **F**.

In this chapter, however, we shall omit 'truth-table' for brevity. An invalid argument thus has at least one row with all premises true and conclusion false. Truth-tables therefore yield a mechanical cut-and-dried test for both validity and invalidity.

Now let us compare argument (I) with the following:

Transylvania will not negotiate unless both the Etruscans pull back their missiles and the Ruritanians stop the flow of weapons to the rebels. However, either the Etruscans will not pull back their missiles or the Ruritanians will not stop the flow of weapons. Therefore Transylvania will not negotiate.

Using fairly obvious abbreviations, this becomes:

(III)   ~N v (E & R)
    ~E v ~R
    ∴ ~N

The same truth-table may be used to show validity here as was used for (I). The reason, of course, is that though arguments (I) and (III) differ in content — in their component sentences — they nonetheless share the same *form*. It should be obvious that *any* argument of this form is valid, inasmuch as the truth-values of the component sentence letters must correspond to one of the truth-value assignments listed in the rows of the truth-table. The table, then, provides a validity test for any argument of this form, not just a particular case like (I) and (III).

We may display the common form of (I) and (III) as a sequence of sentence forms corresponding to their premises and conclusion, in which each variable is paired with a distinct sentence or sentence letter.

(A)   ~p v (q & r)
      ~q v ~r
      ∴ ~p

This array of forms gives the skeletal structure of arguments (I) and (III) since the variables indicate positions within the structure where sentences or sentence letters may be inserted. We shall call any such sequence of sentence forms the last of which is preceded by '∴' an **argument form**, and (I) and (III)—though not of course (II)—are said to be arguments **of the form** (A) displayed above. More fully, let ϕ be any truth-functional argument and ψ any argument form:

DEFINITION ϕ is **of the form** ψ if and only if ϕ can be obtained from ψ by replacing each sentential variable of ψ with a sentence (simple or compound)—the same sentence for the same variable throughout.

Alternatively, we shall say that arguments (I) and (III) are **derivable by substitution** from (A) through the insertion of sentences or letters for the variables. No matter what sentences or letters replace those of (A), the truth-table shows that the resulting argument must be valid. So the truth-table test is best regarded as one for the validity of an argument form rather than of particular arguments *having* that form. Let us then re-consider the above truth-table, this time using sentential variables:

p	q	r	~p	q & r	~p v (q & r)	~q	~r	~q v ~r
T	T	T	F	T	T	F	F	F
T	T	F	F	F	F	F	T	T
T	F	T	F	F	F	T	F	T
T	F	F	F	F	F	T	T	T
F	T	T	T	T	T	F	F	F
F	T	F	T	F	T	F	T	T
F	F	T	T	F	T	T	F	T
F	F	F	T	F	T	T	T	T

This table establishes the validity of argument form (A), and hence of (I), (III), and *any* further argument of that form. A truth-table with variables thus exhibits a measure of generality not shared by tables with sentence letters. Whereas our earlier truth-table established one argument's validity by listing all truth-value combinations its component sentence letters could have, the present table establishes the validity of an argument form—hence any argument of that form—by listing all truth-value combinations that *any* set of three sentences may have. For the practical end of determining validity, it is immaterial whether the truth-table uses letters or variables, but the foregoing considerations should make it clear that validity pertains to the form of an argument, not its content. An argument is therefore valid when it is derivable from a valid argument form by substitution of component sentences or letters for the variables of that form.

Now, a given argument is generally derivable from more than one argument form. Argument (I), for example, is derivable not only from (A) but from

(B) *p* v *q*
     r v *s*
     ∴ *p*

by substitution of '~*S*' for '*p*', '*L* & *I*' for '*q*', '~*L*' for '*r*', and '~*I*' for '*s*'. But a 16-row truth-table will show that argument form (B) is *invalid*. Thus, though a valid argument must be derivable from a valid argument form, this by no means precludes its being derivable from invalid forms as well. What is important, however, is that *only* valid arguments are derivable from valid forms; no matter what substitutions are made for the variables of a valid form—and no matter whether single letters or compounds are substituted—the resulting argument is valid. So the following argument is valid because like (I) and (III) it is derivable from the valid form (A):

(IV) ~(*A* & *B*) v [*C* & (*B* v *D*)]
      ~*C* v ~(*B* v *D*)
      ∴ ~(*A* & *B*).

Here the complex substitutions are '*A* & *B*' for '*p*', '~*C*' for '*q*', and '*B* v *D*' for '*r*'. Summarizing:

(a) To be valid, an argument must be derivable from at least one valid argument form.

(b) *Only* valid arguments are derivable from valid argument forms.

(c) But it is not the case that valid arguments are derivable *only* from valid argument forms—invalid forms yield them too.

Conversely, it follows from (b) that invalid arguments are derivable only from invalid forms. Thus argument (II) is invalid inasmuch as *all* forms from which it results through substitution are invalid.[7]

It will be helpful here to establish via truth-tables the validity of four simple and very fundamental argument forms containing '&', 'v', and '~'. They are associated with the rules of inference commonly known as **Disjunctive Syllogism**, **Addition**, **Simplification**, and **Conjunction**:

Disjunctive Syllogism: $p \lor q$

  $\sim p$

  $\therefore q$

Addition: p

  $\therefore p \lor q$

Simplification: p & q

  $\therefore p$

Conjunction: p

  q

  $\therefore p$ & q

The following truth-table establishes the validity of all four:

p	q	$\sim p$	$p \lor q$	p & q
T	T	F	T	T
T	F	F	T	F
F	T	T	T	F
F	F	T	F	F

Tautologies and contradictions provide very special cases of validity. Since an argument is valid just in case no row of its truth-table yields true premises and false conclusion, it follows that any argument with a tautology as conclusion is valid. That is, *a tautology is implied by any sentence whatever*. The argument form

p

$\therefore q \lor \sim q$

---

[7]The following argument of course is also derivable from (A):

$\sim A \lor (A$ & $A)$

$\sim A \lor \sim A$

$\therefore \sim A.$

While such an argument is of little practical interest, it is nonetheless of form (A) through substitution of '*A*' for each variable. There is nothing wrong with substituting the same sentence for distinct variables — the practice is no different from deriving '$3+3 = 3+3$' from '$x+y = y+x$'.

is thus valid no matter how irrelevant the content of the premise may be to the conclusion. For since there can be no assignment of truth-values to make the conclusion false, there can be no case of true premises and false conclusion. By the same token, since a contradiction comes out false upon all assignments of truth-values, it follows that any argument with a contradiction as premise is valid. In other words, *a contradiction implies any sentence whatever*. Hence the following argument form is valid:

p & ~*p*
∴ *q*

But since the premise can never be true, it would of course be absurd for anyone to propound an argument of this form. Even so, complex arguments sometimes have inconsistent premises, though a cursory examination will not reveal it. A set of premises is said to be **inconsistent** just in case there is no assignment of truth-values to their component letters which makes *all* of them true together. Therefore if we were to form the conjunction of these premises, the resulting expression would clearly have a truth-table column containing nothing but **F**'s. So any argument whose premises are inconsistent will be valid in the same curious way as the above example. In the terminology of the first chapter, such arguments are (trivially) valid, but they cannot possibly be *sound*. It is for this reason that they are in practice useless, although it has happened more than once that a careless protagonist has unwittingly put forth such an argument. The following argument form is valid, though the inconsistency of its premises is far from obvious; the construction of its truth-table is left as an exercise.

p v (~*q* & *r*)
q v ~*p*
~(*p* v *r*)
∴ *p* & *q*

EXERCISES

I.   Test each of the following for validity using truth-tables:
1.   ~*A* v (*B* & *A*)
    ~*B*
    ∴ ~*A*
2.   ~*B* v (*D* & ~*R*)
    R v ~*B*
    ∴ ~*B* v *D*

3. $\sim(E \ \& \ A) \vee R$
    $\sim(R \ \& \ E)$
    $\therefore \sim A \vee \sim R$

4. Either Senator Hominy will not support the incumbent or else
    it is not the case both that the incumbent will seek reelection
    and yet not carry the South. So either the incumbent carries the
    South and Hominy supports him or else he does not seek re-
    election and Hominy does not support him. (H, R, S)

II.   Assume that $\mathcal{T}$ is a sentence such that
      $\mathcal{T} \vee \sim\mathcal{T}$
      $\therefore \mathcal{T}$
      is a truth-table valid argument. Must $\mathcal{T}$ be a tautology? Explain.

III. Assume that $\mathcal{P}$ and $\mathcal{C}$ are sentences such that the argument
      $\mathcal{P}$
      $\therefore \mathcal{C}$
      is truth-table valid. Is $\sim\mathcal{P} \vee \mathcal{C}$ a tautology, a contradiction or
      neither? Explain.

IV.   Where the argument
      $\mathcal{P}_1$
      $\mathcal{P}_2$
      $\vdots$
      $\mathcal{P}_n$
      $\therefore \mathcal{C}$
      is truth-table valid, must
      $\sim(\mathcal{P}_1 \ \& \ \mathcal{P}_2 \ \& \ \ldots \ \& \ \mathcal{P}_n) \vee \mathcal{C}$
      be a tautology? Explain.

Arguments Containing '⊃' and '≡'

There are several fundamental argument forms containing '⊃' whose
validity we shall now consider.[8] They are associated with the rules of
inference traditionally known as **Modus Ponens**, **Modus Tollens**, **Hypothetical Syllogism**, and **Constructive Dilemma**.

Modus Ponens: $p \supset q$
p
$\therefore q$

Modus Tollens: $p \supset q$
$\sim q$
$\therefore \sim p$

Hypothetical Syllogism: $p \supset q$
$q \supset r$
$\therefore p \supset r$

Constructive Dilemma: $p \supset q$
$r \supset s$
$p \vee r$
$\therefore q \vee s$

---

[8]Hereafter for brevity we shall describe sentences of the form '$p \supset q$' simply as "conditionals"—leaving it understood that they are *material* conditionals.

The first two are among the simplest and most pervasive forms of argument. The following truth-table makes their validity evident.

p	q	$\sim p$	$\sim q$	$p \supset q$
T	T	F	F	T
T	F	F	T	F
F	T	T	F	T
F	F	T	T	T

The form corresponding to **Modus Ponens** should not be confused with the invalid form

$p \supset q$
q
$\therefore p$

One who uses this form of argument is said to commit the fallacy of *affirming the consequent*. Its invalidity is shown by the third row of the table; if 'p' is replaced by a false sentence and 'q' by a true one the premises will be true but the conclusion false, as in:

If De Gaulle was the British Prime Minister, he was a politician.
De Gaulle was a politician.
Therefore De Gaulle was the British Prime Minister.

Moreover, the argument form associated with **Modus Tollens** should be distinguished from the invalid form

$p \supset q$
$\sim p$
$\therefore \sim q$

To argue this way is to commit the fallacy of *denying the antecedent*, and its invalidity is also shown by the third row of the above table — and with the same substitutions for 'p' and 'q' as before. The validity of the forms associated with **Hypothetical Syllogism** and **Constructive Dilemma** may be established using truth-tables of eight and sixteen rows respectively. Their construction is left as an exercise for skeptical readers.

We saw that '$p \supset q$' is equivalent to '$\sim p \vee q$', and '$p \equiv q$' to

'$(p \mathbin{\&} q) \vee (\sim p \mathbin{\&} \sim q)$'. Thus the connectives '$\supset$' and '$\equiv$' are by no means essential to our symbolic notation, but their use makes symbolization of many arguments far easier and more natural than relying on just the basic connectives '$\&$', '$\vee$', and '$\sim$'. An obvious example would be:

The President's tax bill will become law if and only if it passes both the House and the Senate. But the Senate will not pass it unless the House does. Consequently the President's tax bill will become law only if the House passes it.

Using obvious letters for the three components, we have:

(V) $B \equiv (H \mathbin{\&} S)$
$\sim S \vee H$
$\therefore B \supset H$

An eight-row truth-table will show the validity of argument (V). However there is a further advantage in using '$\supset$' and '$\equiv$', superfluous though they be in theory. The connective '$\supset$' in particular helps emphasize the intimate connection between arguments and conditionals, and in so doing, provides us with another and more succinct way of characterizing the concept of (truth-table) validity—one which stresses its relation to the notion of truth under all conditions. Let us consider the conditional whose antecedent is the conjunction of the premises of (V) and whose consequent is its conclusion:

(1) $\{[B \equiv (H \mathbin{\&} S)] \mathbin{\&} (\sim S \vee H)\} \supset (B \supset H)$

We shall call (1) the **corresponding conditional** of (V). Every argument has a corresponding conditional, and for an argument of n premises ($n \geqslant 1$)

$\mathscr{P}_1$
$\mathscr{P}_2$
$\vdots$
$\mathscr{P}_n$
$\therefore \mathscr{C}$

its corresponding conditional (omitting inessential parentheses for the moment) may be schematized as

$(\mathscr{P}_1 \mathbin{\&} \mathscr{P}_2 \mathbin{\&} \ldots \mathbin{\&} \mathscr{P}_n) \supset \mathscr{C}$

Now a conjunction is true only when all its conjuncts are, and since the conjuncts correspond to premises, the antecedent of the conditional is true just in case all premises of the argument are true. Moreover, a conditional is true in every case except where the antecedent is true and the consequent false, and notice that this corresponds to the case where the argument has all true premises and false conclusion. Therefore, an argument is invalid just in case its corresponding conditional has at least one **F** in its truth-table column. Since a tautology has *no* **F**'s in its truth-table column, it follows that:

A truth-functional argument (argument form) is *valid* if and only if its corresponding conditional is a tautology (tautologous).

More succinctly, $\mathscr{P}_1$, $\mathscr{P}_2$,..., $\mathscr{P}_n \models \mathscr{C}$ if and only if $\models (\mathscr{P}_1 \ \& \ \mathscr{P}_2 \ \& \ ...$ $\& \ \mathscr{P}_n) \supset \mathscr{C}$. Thus, conditional (1) is a tautology, since its truth-table will have no row in which its antecedent (the conjunction of the premises of (V)) is assigned a **T** and yet its consequent (the conclusion) is assigned **F**. The truth-table is displayed below. Note that instead of a separate column for each of the components, truth-values are assigned to them as they occur within the whole, thus simplifying our work:

B	H	S	{[B	≡	(H	&	S)]	&	(~	S	v	H)}	⊃	(B	⊃	H)
T	T	T	T		T		T F	T		T						
T	T	F	F		F		F T	T		T						
T	F	T	F		F		F F	F		T						
T	F	F	F		F		F T	T		T						
F	T	T	F		T		F F	T		T						
F	T	F	T		F		T T	T		T						
F	F	T	T		F		F F	F		T						
F	F	F	T		F		T T	T		T						

What conditionals are to validity, biconditionals are to equivalence. We saw that two sentence forms are equivalent just in case they have the same truth-value for each assignment of truth-values to their component letters. Now a biconditional is true just in case both components have the same truth-value. Consequently, if we form the biconditional of two equivalent sentences, its truth-table column must be all **T**'s, since for any given row the two sentences are either both assigned a **T** in that row or else both **F**. The biconditional formed from two nonequivalent sentence forms, however, will have at least one **F** in its column, since there must be a row in which the two forms differ in truth-value assignment. Now plainly we can form a biconditional from any pair of sentence forms, and we shall call it their **corresponding biconditional**. We may now refine our notion of equivalence as follows:

Two sentences (sentence forms) are *equivalent* if and only if their corresponding biconditional is a tautology (tautologous).

The following table establishes again the equivalence of '$\sim(A \ \& \ B)$' and '$\sim A \ v \sim B$', this time employing the corresponding biconditional:

A	B		$\sim(A \ \& \ B)$	$\equiv$	$(\sim A \ v \sim B)$	
T	T	F	T	T	F	F F
T	F	T	F	T	F	T T
F	T	T	F	T	T	T F
F	F	T	F	T	T	T T

EXERCISES

I.   Using truth-tables, show the forms corresponding to **Hypothetical Syllogism** and **Constructive Dilemma** to be valid.

II. Put both of the following into symbols and construct their corresponding biconditional. Then determine whether they are equivalent:
   1. Jones will be inducted only if he both passes his physical and is administered the oath of allegiance.
   2. Jones will neither be inducted but fail his physical nor will he not be administered the oath and yet be inducted.

III. Test the following arguments for validity using truth-tables:
   1. $A \supset B$
      $\sim B \ v \sim A$
      $\therefore \sim A$
   2. If the Cubs win, then the other teams have either deteriorated or else sold out to the gamblers. But if they sold out, then the Cubs still will not win. So either the other teams have deteriorated or else the Cubs win. (C, D, G)
   3. If God exists, no animal will suffer extreme pain unless God is unable to prevent it. If God is unable to prevent an animal suffering extreme pain, then He is not omnipotent. It is false that no animal suffers extreme pain. Therefore, either God does not exist or He is not omnipotent. (G, A, P, O)
   4. If either the police or the F.B.I. are not mystified, then Vermicelli was rubbed out either by Mozzarella or Roquefort. But

even though Mozzarella did not rub out Vermicelli, both the F.B.I. and the police are nonetheless mystified. Consequently, Roquefort rubbed out Vermicelli. (*P*, *F*, *M*, *R*)

5. If God wanted to create a universe free from suffering but did not, He is not omnipotent. If God is benevolent, then He wanted to create a perfect universe. God did not create a universe free from suffering. Therefore God is not both omnipotent and benevolent. (*F*, *O*, *B*, *P*)

IV. Let $\mathscr{P}$, $\mathscr{Q}$, and $\mathscr{R}$ be sentences such that $\mathscr{P}, \mathscr{Q} \models \mathscr{R}$. Must it also be the case that $\mathscr{P} \models \mathscr{Q} \supset \mathscr{R}$? What about $\models \mathscr{P} \supset (\mathscr{Q} \supset \mathscr{R})$? Explain.

V. Let the sentential connective '$' be defined as follows:

p	*q*	*p* $ *q*
T	T	T
T	F	T
F	T	F
F	F	T

Which of the following pairs are equivalent?
1. *p* $ *q*; *p* ∨ (~*p* & *q*)
2. *p* $ ~*q*; ~*q* $ *p*
3. *p* $ (~*q* $ *r*); (*p* $ ~*q*) $ *r*
4. (*p* $ *q*) $ *r*; (*p* $ ~*q*) ∨ [(~*p* ∨ *q*) & ~*r*]

*2.4 Another Test: The Method of Assigning Truth-Values

The remarks on corresponding conditionals in the previous section suggest an easier and quicker technique for testing validity than the construction of truth-tables with 16 or more rows. We shall call it **the method of assigning truth-values**, and while it is not always easier than drawing up truth-tables it nonetheless simplifies our work in the majority of cases. The method consists, first, in assigning truth-values to those sentence letters of an argument that suffice to render its conclusion *false*. Then we attempt to make assignments to the remaining letters in such a way as to make all premises true. If such an assignment can be found, the argument is thereby shown invalid; but if for each assignment falsifying the conclusion, *all* attempts to make the premises true end in failure, the argument will then be established as valid.

We need in practice concern ourselves only with those cases where the conclusion is false, inasmuch as validity is established when and only when it has been determined that there is *no* truth-value assignment yielding the true-premise–false-conclusion combination. Thus our new technique of first making the conclusion false, in effect, allows us to consider only those rows of the corresponding truth-table where the conclusion is assigned **F** — ignoring the unproblematic **T** rows.

As might be expected, the truth-value assignment method is used to best advantage in establishing *invalidity*, since only one assignment yielding true premises and false conclusion is required. To show *validity*, one must show that *every* assignment making the conclusion false fails to yield a case of true premises, and this can sometimes be an arduous task. We shall first, however, consider invalid arguments:

(1) $(A \,\&\, B) \supset (\sim C \,\&\, E)$

(2) $D \supset (\sim B \lor \sim C)$

(3) $D \,\&\, A$

$\therefore \sim C$

To falsify the conclusion, we suppose 'C' to be true, and by making assignments to the remaining letters, attempt to render all premises true. Now it is obvious that for (3) to be true, both 'D' and 'A' must be true, so we assign **T** to both. We now take up (2), asking whether it can be true when 'C' and 'D' are true. On these assignments, (2) has a true antecedent, and thus for the whole to be true, its consequent must also be true. Since the consequent is a disjunction with a false disjunct, the other disjunct '$\sim B$' must be true for (2) itself to be true. We therefore assign **F** to 'B'.

Moving now to (1), we shall show invalidity if we can render it true on the basis of the truth-value assignments already made or perhaps, if there are letters as yet unassigned, by making further assignments. In the present case, '*E*' has not yet been assigned a truth-value, but in this instance it is not necessary. For (1) has already been made true by the previous assignments, and thus it does not matter what truth-value '*E*' has. Since '*B*' is false, '*A & B*' is false, thus giving (1) a false antecedent. Therefore (1) is true since a conditional is false *only* when its antecedent is true and consequent false. We have thus found an assignment of truth-values showing the argument to be invalid, and we can display our work as follows:

A	B	C	D	E	(A & B) ⊃ (~C & E)	D ⊃ (~B v ~C)	D & A	~C
T	F	T	T	–	T	T	T	F

Here, we have set out the assignments to the component sentence letters that yield true premises and false conclusion (the dash under '*E*' simply means that it is immaterial whether we assign **T** or **F**). It is easy to see the advantages of this method; had we used a truth-table, it would have required 32 rows.

In thinking our way back through the premises, we implicitly employed certain observations concerning the behavior of the sentential connectives '~', '&', 'v', and '⊃'; for example, that for (2) itself to be true, '*B*' must be false given the truth of '*C*'. The following principles are gleaned from a study of the truth-tables for these connectives, and are helpful in making truth-value assignments:

A negation is false if the negated sentence is true and vice versa.
A conjunction with a false conjunct is false.
A disjunction with a true disjunct is true.
A conditional with a false antecedent is true.
A conditional with a true consequent is true.

Thus, even when assignments have been made only to some of the component letters, there is often much to be inferred concerning the truth-values of the premises and conclusion. However, matters are not always as simple as they were for our sample argument. There, once the conclusion was made false, there was only one assignment that could be made to each of '*A*', '*B*', and '*D*' to render the premises true. But in many cases, there are alternative assignments to a given premise that will make it true, and it may turn out that with only one of them can the rest of the premises be rendered true as well.

The truth-value assignment technique sometimes affords us a quick test of validity too. Consider the elementary argument:

A & B
∴ *B* ∨ ~*A*

Since the conclusion is a disjunction, it can be falsified only by making both disjuncts false, so we assign **F** to '*B*' and **T** to '*A*'. But now we see immediately that the premise cannot be true on this assignment, since both conjuncts must be true and '*B*' of course is not. We may restate the point as follows: for the conclusion to be false '*B*' must be assigned **F**, but for the premise to be true '*B*' must be assigned **T**. Since, therefore, the attempt to obtain a true premise and false conclusion results in a component letter being assigned a **T** at one point and, *per impossible*, an **F** at another, the argument must be valid.

This, then, is how validity is determined: where ϕ is any argument, if for each assignment of truth-values falsifying the conclusion of ϕ, the attempt to make all premises true requires a sentence letter be assigned a **T** at one point and an **F** at another, then ϕ is valid. Armed with this knowledge, then, let us tackle a more challenging argument.

(4) *A* ⊃ *B*

(5) (*A* & *B*) ⊃ *C*

(6) *B* ⊃ (*D* ∨ ~*C*)

 ∴ *A* ⊃ *D*

There is but one assignment making the conclusion false; '*A*' must be assigned **T** and '*D*' assigned **F**. It is now immediately clear that '*B*' must be assigned **T** to make (4) true. However, this makes the antecedent of (5) true, thus requiring that a **T** be assigned to '*C*'. But since '*B*' has been assigned **T** and '*D*' assigned **F**, it is obvious that '*C*' must be assigned **F** in order to make '~*C*'—and hence the consequent of (6)—true. Since incompatible assignments are required of '*C*', the argument is valid. Our work may be summarized as follows:

For the conclusion to be false, '*A*' must be assigned **T** and '*D*' assigned **F**. For (4) to be true, '*B*' must be assigned **T**. But then, for (5) to be true, '*C*' must be assigned **T** and for (6) to be true, '*C*' must be assigned **F**. Therefore the argument is *valid*.

Now if the conclusion of a valid argument can be made false on more than one truth-value assignment, then of course each of these cases will have to be tested in turn. If there are many such assignments – and in addition many further assignments to be made in the premises – the method of assigning truth-values may turn out to be no less cumbersome than the straight truth-table method. On the other hand, should the argument be invalid, the testing comes to an end as soon as a case of true premises and false conclusion emerges – there is obviously no point in going on to consider any further falsifying assignments. There are a great many arguments – both valid and invalid – that are far more easily handled using the present technique. Consider, for example, the forbidding:

(7) $A \supset (\sim B \ \& \ C)$

(8) $D \supset (B \lor \sim C)$

(9) $(\sim E \lor C) \supset D$

(10) $A \supset E$

   $\therefore A \equiv \sim D$

Those who balk at the construction of a 32-row truth-table need instead merely assign truth-values to yield a false conclusion and then test the premises for joint truth. Here, there are two cases where the conclusion will be false: where 'A' and 'D' are both true, and where they are both false.

If 'A' and 'D' are both assigned **T** then the conclusion is false. But for (7) and (8) both to be true, their consequents must be true, and there is no assignment of truth-values to 'B' and to 'C' that will make both '$\sim B \ \& \ C$' and '$B \lor \sim C$' true. If 'B' is assigned **T**, then (8) is true but (7) is not. If 'B' is assigned **F**, then for (7) to be true, 'C' must be assigned **F**, and in that case (8) will be false. So on this assignment, not all premises can be made true.

So far, then, we have established neither validity nor invalidity. But if we can do the same as above for the other falsifying assignment, validity will then be established. However, we cannot do this. For when 'A' and 'D' are both assigned **F**, then (7), (8), and (10) are all true owing to a false antecedent. We need only, then, assign **T** to 'E' and **F** to 'C' to make (9) true, and hence make all premises true:

A	B	C	D	E	$A \supset (\sim B \ \& \ C)$	$D \supset (B \lor \sim C)$	$(\sim E \lor C) \supset D$	$A \supset E$	$A \equiv \sim D$
F	–	F	F	T	T	T	T	T	F

Again, only one case of true premises and false conclusion is needed to show invalidity, so if such a case had arisen on the first assignment, it would not have been necessary to proceed to the second. But to show validity, *all* possible cases with respect to a false conclusion must be traced through (part of the trick, of course, is determining that you *have* exhausted all possibilities), and hence the truth-value assignment technique is shown to its best advantage in establishing invalidity. The technique is most easily applied to arguments whose conclusions admit of falsification in just one or two cases (for instance, '*A*', '*A* ⊃ *B*', '*A* v *B*', '*A* ⊃ (*B* v *C*)'), but for most conclusions admitting of many such cases — as in '(*A* v *B*) ⊃ (*C* & *D*)' — one might well be advised to construct the full truth-table unless it is strongly suspected that one of those cases will quickly lead to all true premises and hence invalidity.

EXERCISES

I. Using the method of assigning truth-values, determine whether the following arguments are valid:

1. $E \supset (J \ \& \ K)$
    $F \supset (E \lor A)$
    $\sim K$
    $\therefore F \supset J$

2. $A \supset (B \ \& \ C)$
    $B \supset (\sim A \ \& \ C)$
    $\therefore \sim A$

3. $A \supset E$
    $B \supset F$
    $\sim E \lor \sim F$
    $\therefore \sim A \lor \sim B$

4. $(A \ \& \ C) \supset F$
    $(B \ \& \ F) \supset [\sim E \supset (D \lor \sim A)]$
    $\therefore (A \ \& \ B) \supset [C \supset (D \lor E)]$

5. $A \supset (B \supset C)$
    $D \supset (B \ \& \ A)$
    $C \supset D$
    $\therefore C \equiv A$

6. $M \lor N$
    $L \lor K$
    $(M \lor L) \supset (N \ \& \ K)$
    $\therefore N \ \& \ K$

7. If Irv does not prove the continuum hypothesis, then although Karl will not gloat, Henry will. However, neither Karl nor Henry will gloat. Consequently, if Irv proves the continuum hypothesis only if Karl gloats, then Irv in fact proves the continuum hypothesis.

8. If the President vetoes the trade bill, Congress will either back him up or else succumb to public pressure. Japan will not raise its tariffs unless the protectionists have their way. If the protectionists have their way and Congress succumbs to public pressure, the world markets will be in disarray. The President did in fact veto the trade bill, so if Japan raises its tariffs, then Congress will not back up the President and the world markets will be in disarray.

9. Milo buys eggs in Sicily at 1 cent per egg and sells them to the distributors in Malta at $4\frac{1}{4}$ cents per egg, and *if* he does both, he clears $3\frac{1}{4}$ cents per egg. If Milo buys eggs in Sicily at 1 cent per egg and buys them back from the distributors in Malta for 7 cents, then he pays a total of 8 cents per egg. If Milo clears $3\frac{1}{4}$ cents per egg and sells them to the commissary for 5 cents, then he takes in $8\frac{1}{4}$ cents per egg. If Milo pays a total of 8 cents per egg and takes in a total of $8\frac{1}{4}$ cents per egg, then he makes a profit. Consequently, if Milo buys back eggs from the distributors in Malta at 7 cents per egg and sells them to the commissary for 5 cents, then he makes a profit.

II. Adapting the method presented in this chapter, determine whether either of the following are tautologies:
1. $[A \supset (B \& C)] \supset \{[(D \& E) \vee \sim B] \supset (D \supset A)\}$
2. $[(A \supset B) \supset A] \supset A$

*2.5 Truth-Functions

A *truth-functional* compound sentence, it will be recalled, is one whose truth-value is determined solely by the truth-values of its component sentences. The concept of truth-functionality is a very important one, and we shall here develop it a bit more fully. Let us note a parallel between truth-functional compound sentences (whether in ordinary discourse or in symbols) and various compound expressions of arithmetic.[9] Just as '$A \vee B$' is a compound containing the sentence letters 'A' and 'B', so '$4 + 3$' is a compound containing the numerals '4' and '3'. In the first case, we have a sentence containing further sentences; in the second, a numerical expression containing other numerical expressions. More importantly, just as the *truth-value* of '$A \vee B$' is determined given the truth-values of 'A' and of 'B', so the *number* signified by '$4 + 3$' is determined given the numbers signified by '4' and '3'. This latter circumstance is customarily described by saying that the number signified by the compound is a *function of* the numbers signified by its components; for as we vary the latter numbers, the identity of the former is correspondingly altered. Thus, given any pair of numbers whatever, if we insert names for them in place of 'x' and 'y' in the expression '$x + y$', the result will be a compound expression signifying a specific number. The expression '$x + y$' thus represents a systematic way of correlating the numbers signified by the resultant compounds with the pairs of numbers signified by the inserted expressions. A **function**, for our purposes, is just such a correlation. The numbers signified by a given pair of numerical expressions inserted in '$x + y$' are said to be **arguments** of the function (called the *addition* function, of course) and the number signified by the compound thus produced the **value** of that function *for* those numbers as arguments. In the present example, the number 7—since it is signified by '$4 + 3$'—is the value of the addition function for the numbers 4 and 3, in that order, as arguments of the function.

The familiar notion of addition, then, may be construed as a function of two arguments, that is, a function which takes a number as its value for an ordered pair of numbers as arguments of the function[10]—such a value being signified by compounds of the form '$x + y$'. Similarly, the notion of *squaring* a number is a function of one argument, for when a single numerical expression is inserted in 'x^2', the resulting compound signifies a

---

[9]Hereafter, we shall use the phrase 'compound sentence' to mean 'truth-functional compound sentence' and restrict our attention to the arithmetic of the integers (whole numbers). Throughout the discussion, the reader must take care not to confuse linguistic *expressions* with what they signify—the expression '3' is a *numeral*, not a *number*; but it signifies or denotes the number 3.

[10]The notion of an **ordered pair** of arguments is important here, for there are functions that yield *different* values for distinct orderings of the same arguments. Subtraction is an example: $6 - 2$ (that is, 4) is a different number from $2 - 6$ (that is, -4).

unique number: the square of the number signified by the inserted expression. The number squared is thus the argument of the function, its square the value for that argument.

Arithmetic functions are distinguished one from another by their having in at least one case *different* values for the same argument(s). Thus, to show that addition and subtraction are distinct functions, it is sufficient to note that the former has the number 6 as its value for the ordered pair of arguments 4 and 2, taken in that order, whereas the latter takes 2 as its value in that circumstance. But the functions represented by '$(x \div 2) + 3$' and '$(x + 6) \div 2$' will obviously have the same value for any argument; no matter which numerical expression is inserted, the resulting compounds always signify the same number. Thus, the functions are one and the same, and this identity is what is asserted in the equation '$(x \div 2) + 3 = (x + 6) \div 2$'.

Now just as there is a functional correlation between the number signified by a compound arithmetic expression and the numbers signified by its components, so is there a functional correlation between the truth-value of a compound sentence and the truth-values of its components. Indeed, it is in virtue of this fact that such sentences are described as "truth-functional"; the sentential connectives '&', 'v', and '~' of our symbolic notation serve as functional signs in much the same fashion as '+' and '−' do in arithmetic discourse. By inserting sentences into '$p \vee q$' we produce a compound sentence whose truth-value is a function of the truth-values of the inserted sentences. Functions that take truth-values both as their values and as their arguments will be called **truth-functions**. The connectives '&' and 'v' each represents a truth-function of *two* arguments.

Now, arithmetic functions and truth-functions differ significantly in the range of items that serve as their arguments, for while there are infinitely many numbers there are only *two* truth-values. This means that for a truth-function of a given number of arguments, we can always specify exactly the number of possible orderings of those arguments—or in other words, **truth-value combinations**—for which the function takes a value. A truth-function of *two* arguments takes a value for exactly four such combinations, which thus correspond to the four rows of the truth-table:

p	q
T	T
T	F
F	T
F	F

Since a conjunction is true only when both conjuncts are, the form '$p \ \& \ q$' expresses that truth-function of two arguments which has the value **T** for the truth-value combination **T–T** and the value **F** for the remaining three: **T–F**, **F–T**, and **F–F**. Therefore this function (the "*and*-function", let us call it) is completely specified by a truth-table column with a **T** in the first row and **F**'s in all the rest. Similarly, a disjunction is true whenever at least one disjunct is, so '*p* v *q*' expresses that truth-function of two arguments which has the value **T** for each of the first three truth-value combinations and the value **F** for the fourth. This "*or*-function", then, is completely specified by a truth-table column having a **T** in every row but the last.

So, unlike arithmetic functions, a truth-function can be completely specified by a table setting out its value for each truth-value combination. This can be done, moreover, regardless of whether it is a function of two arguments or not. One can always determine the number of truth-value combinations for a truth-function of any number of arguments (and thereby, of course, the number of truth-table rows required):

For a truth-function of *n* arguments, the number of truth-value combinations is 2^n.

We have already seen that where $n = 2$, there are 4 (or 2^2) combinations of truth-values. Clearly for a truth-function of just one argument, such as that represented by '∼', that argument can be only **T** or **F** — since there is but one argument, the notion of a truth-value "combination" is redundant here. Thus, any compound sentence with just one component requires a table with 2^1 rows, and where '*p*' represents any such component, we have as before:

p
T
F

A sentence of the form '∼*p*' will be true when the sentence negated is false and vice versa, so '∼*p*' expresses that truth-function of one argument (the "*not*-function", to give it a name) which has the value **T** for the argument **F**, and **F** for the argument **T**. This function is thus specified by a truth-table column with an **F** in the first row and a **T** in the second.

For a truth-function of *three* arguments there will be, according to our formula, 2^3 distinct truth-value combinations for which it takes a value. Hence, compounds with three component sentences require, as we have seen, a truth-table of eight rows. Truth-functions of *four* arguments will have 2^4 truth-value combinations, and so compounds with four components require 16-row truth-tables for the specification of their truth-conditions. Functions of five or more arguments clearly require quite sizable — and cumbersome — truth-tables.

To show that two truth-functions are distinct functions, it suffices to exhibit one truth-value combination for which they take different truth-values as values. In the case of the *and-* and *or*-functions, the second and third rows of their truth-tables provide such combinations.

By the same token, truth-functions are identical when, for every truth-value combination, they yield the same truth-value. Since, for example, the columns for '$\sim(p \lor q)$' and '$\sim p \mathbin{\&} \sim q$' have the same arrangement of **T**'s and **F**'s, sentences derivable from them via the same substitutions are true and false under the same conditions. These forms thus provide different ways of expressing the very same truth-function (the "*neither-nor*–function", as we shall call it). It is that truth-function of two arguments that has the value **T** for the truth-value combination **F–F** and the value **F** for the rest. And as we can see from the truth-table in Section 2.1, no special connective need be introduced to express this function — the job may be done either by '$\sim$' and '&' or else by '$\sim$' and '$\lor$'. *Equivalent* sentence forms, therefore, express one and the same truth-function.

Since there are only two truth-values to serve as arguments for a truth-function, we were able to state a formula by which we could determine for any truth-function the number of truth-value combinations for which it takes a value. We may also determine how many truth-functions there are for a given number of arguments:

The number of truth-functions of *n* arguments is 2^{2^n}.

Consider truth-functions of one argument. Here there are 2^{2^1} or 4 truth-functions, for that is as many distinct arrangements of **T**'s and **F**'s as there are for a two-row truth-table. Thus any sentence form with just one component must express one (and only one) of these four functions. They are specified as follows:

p	$f_1(p)$	$f_2(p)$	$f_3(p)$	$f_4(p)$
T	T	T	F	F
F	T	F	T	F

We have long since introduced the connective '$\sim$' for representing the function listed here as '$f_3(p)$'; $f_3(p)$ is thus the "*not*-function". Note that $f_2(p)$ has the same arrangement of **T**'s and **F**'s as the component 'p' itself; we may thus express $f_2(p)$ simply by 'p'. That is, the truth-value of any sentence is (trivially) a function of itself—a truth-function having the value **T** for the argument **T** and **F** for the argument **F**. The *verum* function $f_1(p)$ has the value **T** for both arguments, and may be expressed by '$p \lor \sim p$'. Similarly, the *falsum* function $f_4(p)$ has **F** as its value for both arguments, and is thus expressible by '$p \mathbin{\&} \sim p$'.

Now there are 2^{2^2} or 16 truth-functions of two arguments, for there are exactly 16 distinct arrangements of **T**'s and **F**'s for a four-row truth-table, including the limiting cases of all **T**'s and all **F**'s. Any compound sentence of two components must therefore express one of these functions:

p	q	$f_1(p, q)$	$f_2(p, q)$	$f_3(p, q)$	$f_4(p, q)$
T	**T**	**T**	**T**	**T**	**T**
T	**F**	**T**	**T**	**T**	**F**
F	**T**	**T**	**T**	**F**	**T**
F	**F**	**T**	**F**	**T**	**T**

p	q	$f_5(p, q)$	$f_6(p, q)$	$f_7(p, q)$	$f_8(p, q)$
T	**T**	**F**	**T**	**T**	**T**
T	**F**	**T**	**T**	**F**	**F**
F	**T**	**T**	**F**	**T**	**F**
F	**F**	**T**	**F**	**F**	**T**

p	q	$f_9(p, q)$	$f_{10}(p, q)$	$f_{11}(p, q)$	$f_{12}(p, q)$
T	**T**	**F**	**F**	**F**	**F**
T	**F**	**T**	**T**	**F**	**F**
F	**T**	**T**	**F**	**T**	**F**
F	**F**	**F**	**T**	**T**	**T**

p	q	$f_{13}(p, q)$	$f_{14}(p, q)$	$f_{15}(p, q)$	$f_{16}(p, q)$
T	**T**	**F**	**F**	**T**	**F**
T	**F**	**F**	**T**	**F**	**F**
F	**T**	**T**	**F**	**F**	**F**
F	**F**	**F**	**F**	**F**	**F**

Many of these are already familiar: $f_2(p, q)$ is of course the *or*-function and expressible by '$p \lor q$'; $f_{15}(p, q)$ is the *and*-function expressible by '$p \mathbin{\&} q$'; the *neither-nor*-function is $f_{12}(p, q)$ and may be expressed by '$\sim p \mathbin{\&} \sim q$' or equivalently by '$\sim(p \lor q)$'. Moreover, $f_5(p, q)$ might be termed the "*not-both*-function" since it is expressed by '$\sim(p \mathbin{\&} q)$' and by '$\sim p \lor \sim q$'. The "*exclusive-or*-function"—which may be expressed in symbols by '$(p \mathbin{\&} \sim q) \lor (\sim p \mathbin{\&} q)$'—is here displayed as $f_9(p, q)$.

Since the *neither-nor*-, *not-both*-, and *exclusive-or*-functions may all be expressed in terms of '&', 'v', and '~', the question naturally arises whether these connectives alone are capable of expressing the remaining two-argument truth-functions as well. The answer is: not only can all of them be expressed, but we can dispense with 'v', using only '&' and '~' ('p v q' is equivalent to '$\sim(\sim p \; \& \sim q)$'). That is, having introduced '~' for $f_3(p)$ and '&' for $f_{15}(p, q)$, all the remaining 18 functions of one and two arguments may be expressed by sentence forms containing, in addition to the necessary parentheses and sentential variables, one or more occurrences of *just* these connectives. What is more surprising is that '~' and '&' can do the same for *all* truth-functions whatever! Any truth-function of *any* number of arguments may be expressed by a sentence form whose only connectives are '~' and '&'. To fully appreciate the extent of this claim, one should keep in mind that there are 2^{2^3} or 256 truth-functions of three arguments, 2^{2^4} or 65,536 truth-functions of four arguments, and so forth.

To bear out this contention, let us first consider some examples. The function $f_{13}(p, q)$ has the value **T** for just one truth-value combination, that given in row three. In this row, 'p' is assigned an **F** but 'q' a **T**. Since a conjunction is true only when both conjuncts are, '$\sim p \; \& \; q$' takes a **T** in *just* this row and so expresses $f_{13}(p, q)$. More generally, any truth-function specified by a truth-table column having a **T** in just *one* row can be expressed by a sentence form in which the simple components are conjoined together, but those to which **F** is assigned in that row are negated. So $f_{13}(p, q)$ may be expressed solely in terms of '~' and '&' by '$\sim p \; \& \; q$', and similarly $f_{14}(p, q)$ may be expressed by '$p \; \& \sim q$'.

But now we must consider functions that have the value **T** for more than one case. The function $f_{10}(p, q)$ has the value **T** for each of the truth-value combinations given in rows 2 and 4. Since it has the value **T** in just these cases, then, it can be expressed by '$(p \; \& \sim q)$ v $(\sim p \; \& \sim q)$'. For this expression is assigned a **T** just in case either '$p \; \& \sim q$' or '$\sim p \; \& \sim q$' is assigned a **T**; that is, just in case 'p' is assigned a **T** and 'q' an **F** (row 2) or else 'p' and 'q' are both assigned **F**'s (row 4). By the same technique $f_3(p, q)$ is expressible—dropping unneeded parentheses for the moment—by '$(p \; \& \; q)$ v $(p \; \& \sim q)$ v $(\sim p \; \& \sim q)$'. In general, a truth-function specified by a truth-table column having **T**'s in two or more rows is expressible by a disjunction each of whose disjuncts is a conjunction of the type described above. If there are k rows with a **T**, there will be exactly k conjunctions disjoined together. The next step is to eliminate the 'v', and this is easily accomplished since we already know that 'p v q' may be equivalently expressed as '$\sim(\sim p \; \& \sim q)$'. Applying this knowledge to our examples, $f_{10}(p, q)$ is expressible by '$\sim[\sim(p \; \& \sim q) \; \& \sim(\sim p \; \& \sim q)]$' and $f_3(p, q)$ by '$\sim[\sim(p \; \& \; q) \; \& \sim(p \; \& \sim q) \; \& \sim(\sim p \; \& \sim q)]$'. So functions having the value **T** in two or more cases can also be expressed solely in terms of '~' and '&'.

The same pattern of reasoning may be extended to functions of three or more arguments. Two such functions are:

p	q	r	$f_j(p, q, r)$	$f_k(p, q, r)$
T	T	T	F	T
T	T	F	F	F
T	F	T	T	T
T	F	F	F	F
F	T	T	F	F
F	T	F	F	T
F	F	T	F	F
F	F	F	F	F

The function $f_j(p, q, r)$ has the value **T** for just one truth-value combination—that given in row three. It thus has the value **T** in just the row in which '$p \mathrel{\&} \sim q \mathrel{\&} r$' is assigned a **T**, and is therefore expressible by that form. For $f_k(p, q, r)$ we again form such a conjunction for each of the three rows in which a **T** occurs and then disjoin them, obtaining '$(p \mathrel{\&} q \mathrel{\&} r) \lor (p \mathrel{\&} \sim q \mathrel{\&} r) \lor (\sim p \mathrel{\&} q \mathrel{\&} \sim r)$'. Eliminating '$\lor$' as before, we derive

$$\sim[\sim(p \mathrel{\&} q \mathrel{\&} r) \mathrel{\&} \sim(p \mathrel{\&} \sim q \mathrel{\&} r) \mathrel{\&} \sim(\sim p \mathrel{\&} q \mathrel{\&} \sim r)]$$

which contains only the connectives '&' and '$\sim$'. There are simpler expressions for this function, such as '$(p \mathrel{\&} r) \lor (\sim p \mathrel{\&} q \mathrel{\&} \sim r)$', but the important consideration is that the technique sketched here will *always* give the required sentence form if there is at least one **T** in the column.

Plainly the technique may be extended to functions of four or more arguments as well, the only complication being the size of the truth-tables required. For any number n of sentential variables: (1) construct the appropriate 2^n row truth-table; (2) for any arbitrary column containing k number of **T**'s ($1 \leq k \leq 2^n$), produce a conjunction of the n components for each **T** row, negating each component to which **F** is assigned in that row; (3) if $k > 1$, disjoin them and eventually eliminate '$\lor$'. We thus have a routine by which we can always *find* an expression whose only connectives are '$\sim$' and '&' to represent any randomly constructed column of **T**'s and **F**'s of any truth-table—in effect, to express any truth-function of any number of arguments. The only case left to consider is where only **F**'s occur in the column, as in $f_4(p)$ and $f_{16}(p, q)$. Now any sentence of the form '$p \mathrel{\&} \sim p$' must be false no matter what the components' truth-values are, and so $f_4(p)$ may be expressed as '$p \& \sim p$'. To express $f_{16}(p, q)$, we simply write '$p \& \sim p \& q$'. Our routine thus breaks down into three cases:

(a) exactly one **T** in the column: conjoin the components, negating those to which an **F** is assigned in the **T** row;

(b) more than one **T** in the column: produce the conjunctions for each **T** row described in (a), negate each and conjoin them, then negate the whole;

(c) no **T**'s in the column: conjoin all components along with the negation of any one of them.

Thus '$\sim$' and '$\&$' form a **truth-functionally complete** set of connectives — they are capable of expressing all of our truth-functional discourse. But as some readers may already have suspected, '$\&$' can be eliminated in favor of 'v'. Since '$p \& q$' is equivalent to '$\sim(\sim p \ v \sim q)$', the connectives '$\sim$' and 'v' form a complete set too. However, '$\&$' and 'v' form an incomplete set, since there are truth-functions which cannot be represented by expressions containing only those connectives. We leave it to the reader to show that they cannot alone express the *neither-nor*-function.

EXERCISES

I. Produce a sentence form containing 'p', 'q', and 'r' whose connectives are '$\sim$' and '$\&$', and which is assigned a **T** *only* when 'q' is assigned a **T** and 'r', an **F**.

II. Let us introduce the connective '$|$' (the *stroke*) for the *not-both*-function, $f_5(p, q)$:

p	q	$p \mid q$
T	T	F
T	F	T
F	T	T
F	F	T

Surprisingly, this connective, *by itself*, can express all truth-functions whatever! Since we have already shown that '$\sim$' and '$\&$' form a complete set, we can establish the completeness of '$|$' if we can produce expressions equivalent to '$\sim p$' and to '$p \& q$' whose only sentential connective is '$|$'. Find such expressions, using a truth-table to justify your answers.

Hint: for negation, try to find an expression built up from 'p' and '$|$' which has **F**'s in the first two rows and **T**'s in the last two.

III. We represent the *neither-nor*-function, $f_{12}(p, q)$, by '$\downarrow$' (the *dagger*):

p	q	$p \downarrow q$
T	T	F
T	F	F
F	T	F
F	F	T

This connective can also express all truth-functions. Show the completeness of '$\downarrow$' in the same manner you did for '$|$' in the preceding exercise. Then express each of the following in terms of '$\downarrow$': $f_2(p, q)$; $f_4(p, q)$; $f_5(p, q)$.

*IV. There are two robots, Boris and Bela. If you utter a true sentence, Boris always responds "True!", and if you utter a falsehood, he responds "False!" Bela, on the other hand, is a fiend who replies "False!" upon hearing a true sentence, "True!" upon hearing a falsehood. You approach a robot wishing to find out which fork in the road ahead will take you to East Grubbsville, your destination. However, you do not know if the robot in front of you is Boris or Bela. You are allowed to utter just *two* (declarative) sentences to the robot, and must determine from his answers which road will take you to East Grubbsville. What sentences should you utter? Suppose you are allowed to utter just *one* sentence—find one which will do the job!

CHAPTER THREE

SENTENTIAL LOGIC
Deductive Methods

3.1 Rules of Inference and Replacement for '&', 'v', and '∼'

It is not hard to see that the argument

Either the number two is composite or else the number two is prime.	C v P
It is not the case that the number two is composite.	∼C
Therefore the number two is prime.	∴ P

is valid: if its premises are both true, then its conclusion has to be true too. In fact it's obvious that any argument of the form

$p \lor q$
$\sim p$
$\therefore q$

will be valid in that sense, for if a disjunction is true but its first disjunct is not then its second disjunct must be.

We have an equally easy time with examples like

God is omnipotent and benevolent.	O & B
Therefore God is omnipotent.	∴ O

and

God is omnipotent.	O
Therefore God is omnipotent or omniscient.	∴ O v K

These arguments, as well as all others of the forms

p & q
∴p

and

p
∴p v q

are clearly valid: whenever a conjunction is true its first conjunct must be, and whenever the first disjunct of a disjunction is true the whole disjunction must be.

Things get a little more complicated with

God is omnipotent and benevolent.	O & B
Therefore God is either omnipotent or omniscient.	∴O v K

but this argument can still be handled fairly easily if we combine two of the things we've already noticed: it follows from the premise that God is omnipotent, and from that it follows in turn that God is either omnipotent or omniscient.

When, however, we come to such a complicated argument as the following, a bit more work is required:

Either our senses are wholly reliable, or else our sense of sight sometimes deceives us, our sense of touch sometimes misleads us, and our sense of hearing sometimes gives us false impressions.	R v $[(S$ & $T)$ & $H]$
It is not the case that our senses are wholly reliable.	~R
Therefore either our sense of sight sometimes deceives us or else our sense of sight always deceives us.	∴S v A

From the two premises of this argument it follows that

Our sense of sight sometimes deceives us, our sense of touch sometimes misleads us, and our sense of hearing sometimes gives us false impressions.	$(S$ & $T)$ & H

The intermediate argument from those premises to that conclusion is an argument of the form

p v q
~p
∴q

(substituting '*R*' for '*p*' and '(*S* & *T*) & *H*' for '*q*'), and it is clear, as we've already noticed, that any argument of that form is a valid one.

But of course whenever

Our sense of sight sometimes deceives us, our sense of touch sometimes misleads us, and our sense of hearing sometimes gives us false impressions	(*S* & *T*) & *H*

is true,

Our sense of sight sometimes deceives us and our sense of touch sometimes misleads us	*S* & *T*

will be true, too — any argument of the form

$p \& q$
$\therefore p,$

we've noticed, is valid.

For the same reason, when

Our sense of sight sometimes deceives us and our sense of touch sometimes misleads us	*S* & *T*

is true,

Our sense of sight sometimes deceives us	*S*

must also be true; and from the latter we can infer

Either our sense of sight sometimes deceives us or else our sense of sight always deceives us,	*S* v *A*

since all arguments of the form

p
$\therefore p \text{ v } q$

are valid.

We've shown that the conclusion of our original argument follows from the premises of that argument by *deducing* it from them, taking care at each stage of the game to make *only* inferences that could be justified by reference to certain elementary forms of argument. Since each of the argument forms we relied upon along the way can take us from true sentences only to other true sentences, we could be sure at each point in the deduction that the sentence being inferred at that point, from preceding sentences, had to be true when those preceding sentences were. At the end, then, we could be sure that the conclusion of our original argument had to be true if its premises were, and thus be sure that our original argument was valid.

The whole **deduction** (as we shall call it) can be written up so as to record our train of thought by listing in order the chain of inferences we've put together:

R v $[(S \& T) \& H]$
$\sim R$
$(S \& T) \& H$
$S \& T$
S
S v A

To clarify the whole thing we can number the various lines of our deduction and note, to the right, what is involved at each point. We began with two premises and wanted to deduce a certain conclusion from them; so we can begin by writing

1. R v $[(S \& T) \& H]$ Premise

2. $\sim R$ Premise   DEDUCE: S v A

The first thing we noticed was that '$(S \& T) \& H$' followed from these two premises, so we continue:

3. $(S \& T) \& H$ 1, 2

Since the rule of inference we are relying on here — the rule that permits one to infer from a disjunction and the negation of its first disjunct its second disjunct — has a name, **Disjunctive Syllogism** (usually abbreviated **DS**), we can note that too:

3. $(S \& T) \& H$ 1, 2 DS

We next inferred '*S* & *T*' from (3) on the grounds that any argument whose premise is of the form '*p* & *q*' and whose conclusion is the corresponding sentence '*p*' is valid. The rule that permits one to infer from any conjunction its left conjunct is usually called **Simplification** (abbreviated **Simp**) so we write, as our fourth step:

4. *S* & *T*   3 Simp

We then noticed that '*S*' follows from (4) on the same grounds:

5. *S*   4 Simp

For our final step we recalled that any argument whose conclusion is of the form '*p* v *q*' and whose premise is the corresponding sentence '*p*' is valid, and so deduced '*S* v *A*' from (5). The rule that permits one to infer from any sentence any disjunction whose first disjunct is that sentence is called **Addition** (abbreviated **Add**), so we can finish writing up our deduction by writing:

6. *S* v *A*   5 Add

The completed deduction looks like this:

1. *R* v [(*S* & *T*) & *H*]   Premise

2. ~*R*   Premise   DEDUCE: *S* v *A*

3. (*S* & *T*) & *H*   1, 2 DS

4. *S* & *T*   3 Simp

5. *S*   4 Simp

6. *S* v *A*   5 Add

To give ourselves some range, we begin with four rules of inference associated with four valid forms of argument, along with a special rule allowing us to introduce sentences of the form '*p* v ~*p*' into deductions whenever we want:

Rules of Inference	*Justification*

Conjunction (Conj)

In any deduction, one may infer from two previously obtained sentences

$$p$$

and

$$q$$

the corresponding sentence

$$p \,\&\, q.$$

Any argument of the form

$$p$$
$$q$$
$$\therefore p \,\&\, q$$

is valid, for if each of a pair of sentences is true then the conjunction of the two must also be true.

Simplification (Simp)

In any deduction, one may infer from a previously obtained sentence of the form

$$p \,\&\, q$$

the corresponding sentence

$$p.$$

Any argument of the form

$$p \,\&\, q$$
$$\therefore p$$

is valid, for if a conjunction is true then its first conjunct must also be true.

Addition (Add)

In any deduction, one may infer from a previously obtained sentence

$$p$$

any sentence of the form

$$p \lor q.$$

Any argument of the form

$$p$$
$$\therefore p \lor q$$

is valid, for if the first disjunct of a disjunction is true then the whole disjunction must also be true.

Disjunctive Syllogism (DS)

In any deduction, one may infer from two previously obtained sentences of the forms

$$p \lor q$$

and

$$\sim p$$

the corresponding sentence

$$q.$$

Any argument of the form

$$p \lor q$$
$$\sim p$$
$$\therefore q$$

is valid, for if a disjunction is true but its first disjunct is not then its second disjunct must be.

**Excluded-Middle Introduction
(E-M I)**

At any point in a deduction, one may introduce any sentence of the form

$$p \text{ v } \sim p.$$

No matter what sentence is substituted for 'p', the disjunction

$$p \text{ v } \sim p$$

must be true, for if the first disjunct is not true the second is.

EXERCISES

I. Deduce the conclusions of each of the following arguments from their premises:

1. $A \text{ v } B$
$\sim A \text{ \& } C$
$\therefore (B \text{ v } \sim D) \text{ \& } (D \text{ v } \sim D)$

2. $\sim B$
$[B \text{ v } (\sim A \text{ v } C)] \text{ \& } (A \text{ v } B)$
$\therefore (\sim A \text{ v } C) \text{ \& } (\sim B \text{ v } C)$

3. C
$\therefore C \text{ \& } [(D \text{ v } \sim D) \text{ v } (E \text{ v } \sim E)]$

4. $C \text{ \& } D$
$\therefore [C \text{ \& } (D \text{ v } \sim D)] \text{ v } (E \text{ v } \sim E)$

5. $(C \text{ \& } D) \text{ \& } (E \text{ \& } F)$
$\therefore (C \text{ v } B) \text{ v } \sim B$

6. $N \text{ v } (S \text{ v } \sim A)$
$\sim N \text{ \& } \sim P$
$\sim S \text{ \& } \sim Q$
$\therefore \sim N \text{ \& } (\sim S \text{ \& } \sim A)$

II. Make up an argument such that the deduction of its conclusion from its premises requires that each of our five rules be used at least twice.

III. Make up an argument such that the deduction of its conclusion from its premises requires that each of our five rules be used exactly twice.

* IV. Let's use the symbol '*' temporarily (for this exercise only) as an abbreviation for the English expression '... and possibly ---'. Then, for example, we can abbreviate the sentence 'Raquel is coming to the party, and possibly Paula is too' by writing '$R * P$'. Which of the following arguments are valid?

$R * P$ R
$\therefore R$ P
         $\therefore R * P$

Is every argument of the form

$p * q$
$\therefore p$

valid? How about arguments of the form

p
q
$\therefore p * q$?

*V.   Make up an argument that is valid but that *cannot* be handled using just our five rules, thus establishing that we need to add a few more rules to our list. (One of the difficulties, of course, is finding a way to show that no combination of our rules will suffice to allow the deduction of your argument's conclusion from its premises.)

Rules of Replacement

Let's examine in detail what happens when we try to treat the clearly valid argument

Fabre's insect studies were either useless, significant, or not unhelpful. $U \lor (S \lor \sim\sim H)$

Fabre's insect studies were not useless. $\sim U$

Therefore Fabre's insect studies were either significant or helpful. $\therefore S \lor H$

From the two premises we can obviously deduce

$S \lor \sim\sim H$

by **DS**; and it is clear that

$S \lor H$

follows from the former in the sense that '$S \lor \sim\sim H$' cannot be true unless '$S \lor H$' is also. But it turns out that our five rules do not permit us to infer the latter from the former.[1] What we need is a rule that somehow reflects the fact that in contexts such as this one a sentence and its double negation are *interchangeable*.

---

[1]A good exercise (see Exercise V above) would be to try to give some convincing justification for this claim.

Notice, by the way, that it will *not* do to add to our list a rule like this:

At any point in any deduction, one may infer from any previously obtained sentence

p

the corresponding sentence of the form

$\sim\sim p$;

and similarly from

$\sim\sim p$

one may infer

p.

There wouldn't be anything *wrong* with adding such a rule, of course; any argument of the form

p
$\therefore \sim\sim p$

as well as any argument of the form

$\sim\sim p$
$\therefore p$

is valid. It's just that such a rule would *not* help us with the problem confronting us here. We're working with the sentence 'S v $\sim\sim H$', which is clearly not of the form '$\sim\sim p$'. We want to go *inside* that sentence and tinker with just a part of it (in fact, with a *part* of the form '$\sim\sim p$'). What we need is a rule that will let us go inside lines of deductions, replacing certain of their parts. Where 'p' represents any sentence, let us use '...p...' to represent any sentence in which it occurs; then the rule we need could be formulated along the following lines:

Double Negation (DN)

In any deduction, one may infer from any previously obtained sentence

...p...

containing an occurrence of some sentence, 'p', any corresponding sentence

...$\sim\sim p$...

provided that the latter is *just like* the former *except* for containing an occurrence of the corresponding sentence

   $\sim\sim p$

in exactly one place where the former contained an occurrence of 'p'; and similarly, from

   $\ldots\sim\sim p\ldots$

one may infer

   $\ldots p\ldots$

To avoid such a long statement of this rule, we'll simply write

Double Negation (DN)
$p :: \sim\sim p$

intending the odd mark '::' to be read as something like '…can replace or be replaced by, at any point in any line of a deduction…'.

We also need some other rules of a similar sort, for example one that will allow us to replace in any line of any deduction a sentence of the form

$p \,\&\, q$

with the corresponding sentence of the form

$q \,\&\, p.$

The following longish set of rules has proved useful enough in the past to justify its length.

Rules of Replacement[2]

Commutation (Com)
$p \,\&\, q :: q \,\&\, p$
$p \vee q :: q \vee p$

Association (Assoc)
$p \,\&\, (q \,\&\, r) :: (p \,\&\, q) \,\&\, r$
$p \vee (q \vee r) :: (p \vee q) \vee r$

---

[2]Intuitive justifications for each of these rules can be supplied by the reader, using those given in Section 1.3 for **De Morgan** as a guide. Going left to right on the first of the **Distribution** rules, for example, we can think along the following lines: let 'p', 'q', and 'r' represent any sentences. Assume that 'p' is true and that the disjunction 'either q or r' is also true. Now

Distribution (Dist)
$p \;\&\; (q \lor r) :: (p \;\&\; q) \lor (p \;\&\; r)$
$p \lor (q \;\&\; r) :: (p \lor q) \;\&\; (p \lor r)$

Double Negation (DN)
$p :: {\sim}{\sim}p$

De Morgan (De M)
${\sim}p \;\&\; {\sim}q :: {\sim}(p \lor q)$
${\sim}p \lor {\sim}q :: {\sim}(p \;\&\; q)$

Using our full set of ten rules we can now show a number of arguments valid by the simple expedient of deducing their conclusions from their premises. Consider, for example, the following argument:

Either the disjunction you're committed to is true, or else its left and right disjuncts are both false.	$T \lor (L \;\&\; R)$
Therefore either the disjunction you're committed to is true, or else its left or right disjunct is false.	$\therefore T \lor (L \lor R)$

None of our ten rules permits us to move directly from the premise of this argument to its conclusion. But by using several of them, one at a time[3], we can get there:

1. $T \lor (L \;\&\; R)$ Premise   DEDUCE: $T \lor (L \lor R)$

2. $(T \lor L) \;\&\; (T \lor R)$ 1 Dist

3. $T \lor L$ 2 Simp

4. $(T \lor L) \lor R$ 3 Add

5. $T \lor (L \lor R)$ 4 Assoc

---

if 'q' is true, then 'p' and 'q' are both true and so 'either p and q, or p and r' will be true because its first disjunct is. On the other hand, if 'q' is *not* true then 'r' must be (since 'either q or r' is); in that case both 'p' and 'r' are true, so again 'either p and q, or p and r' will be true, this time because its *second* disjunct is.

More satisfactory justifications can be supplied by truth-tables — see Section 2.1.

[3]**Simp** only permits us to infer the *left* conjunct of a conjunction, not the right one. So long as we permit ourselves to use but one rule per line, then, there is no way to move from '$A \;\&\; B$' to 'B' in a single step. Instead we must first apply **Com** to '$A \;\&\; B$' to get '$B \;\&\; A$', and **Simp** will then deliver 'B'. Similar problems come up when we want to use **DS** with 'C' and '${\sim}C \lor D$'; 'C' is not the negation of the left disjunct of '${\sim}C \lor D$', so we must use **DN** to get '${\sim}{\sim}C$' before **DS** can be applied. Of course we might justifiably add rules to our system that would permit us to move directly from '$A \;\&\; B$' to 'B' and from 'C' and '${\sim}C \lor D$' to 'D', but with **Com** and **DN** available they would be more trouble than they're worth.

For now we'll put up with minor inconveniences of the above sort. But later on (see Section 3.4), we shall allow ourselves to combine **Com** or **DN** with any other rule as a single step.

We first notice that the premise gives us two pieces of information, really—the two conjuncts of (2). The first of these, (3), is enough to assure us of (4), and we need only rearrange parentheses in (4) to get the conclusion we're after.

Since there are infinitely many valid arguments of distinct forms and we're trying to get by on just ten rules, our deductions will sometimes have to be rather lengthy. But it should be noted that we shall often have available several different ways of deducing the conclusion of an argument from its premises using our ten rules, and the reader should realize that facility in discovering short deductions can be a great timesaver. For example, the most direct deduction of '(~A & ~B) & D' from the premises '~A & (~B & ~C)' and '(A v B) v D' is, perhaps, the following one:

1. ~A & (~B & ~C)	Premise	
2. (A v B) v D	Premise	DEDUCE: (~A & ~B) & D
3. ~A	1 Simp	
4. A v (B v D)	2 Assoc	
5. B v D	4, 3 DS	
6. (~B & ~C) & ~A	1 Com	
7. ~B & ~C	6 Simp	
8. ~B	7 Simp	
9. D	5, 8 DS	
10. ~A & ~B	3, 8 Conj	
11. (~A & ~B) & D	10, 9 Conj	

We simply take (1) apart to get (3) and later (8); we use (3) and (8) on (2) until we get (9); and then we stick (3), (8), and (9) together to get our conclusion. But ingenious use of our rules, including some not-so-obvious moves, can cut down the length considerably:

1. ~A & (~B & ~C)	Premise	
2. (A v B) v D	Premise	DEDUCE: (~A & ~B) & D
3. (~A & ~B) & ~C	1 Assoc	
4. ~A & ~B	3 Simp	
5. ~(A v B)	4 De M	
6. D	2, 5 DS	
7. (~A & ~B) & D	4, 6 Conj	

Sometimes, of course, such short deductions cannot be found and one has to resign himself to finding one fifteen or twenty lines long. But readers will generally find that the more familiar they are with our ten rules and the various ways they can work in combination, the easier it will be to discover deductions of reasonable length. And tricks discovered in the process of working through easier problems often turn out to be helpful when one comes to the harder ones. The following deduction, for example, is fairly easy to discover:

1. A v $(B$ & $\sim B)$ Premise   DEDUCE: A

2. $\sim B$ v $\sim\sim B$ E-M I

3. $\sim(B$ & $\sim B)$ 2 De M

4. $(B$ & $\sim B)$ v A 1 Com

5. A 3, 4 DS

The second conjunct of (1) cannot possibly be true; in fact, we can establish its negation via **E-M I** and **De M**. Commutation of (1) then sets up a finish with **DS**. And now when we come to the deduction beginning

1. B v B Premise   DEDUCE: B

prospects are not so bleak as they might have been. Of course, since we have but one premise, few of our rules apply. Even those that do scarcely seem to lead anywhere — we could obtain 'B v $\sim\sim B$' by **DN** or '$(B$ v $B)$ v $\sim B$' by **Add**, but neither gets us closer to the plain 'B'. Suppose, however, that we could deduce 'B v $(B$ & $\sim B)$' from (1). Then the lesson learned on the preceding deduction would provide the rest — we'd have only to use **E-M I** and **De M** to get '$\sim(B$ & $\sim B)$', and then commute 'B v $(B$ & $\sim B)$' and do **DS**. The only remaining problem, then, is to find a way to deduce 'B v $(B$ & $\sim B)$' from (1); and here our experience with **Dist** on the earlier examples comes into play. 'B v $(B$ & $\sim B)$' could be obtained by **Dist** from '$(B$ v $B)$ & $(B$ v $\sim B)$' — the left conjunct of the latter is our premise and the right conjunct easily obtained by **E-M I**!

EXERCISES

Symbolize each of the following arguments, using the letters suggested. Then deduce their conclusions from their premises.

1. The number two is not both prime and composite. The number two is prime. Therefore the number two is not composite. (P, C)

2. The number two is not composite. Therefore the number two is not both composite and prime. (C, P)

3. Either the first witness told the truth or the third witness did not. The first witness did not tell the truth, but either the second witness or the third one did. Therefore the second witness told the truth. (F, S, T)

4. Either the first witness or the second witness lied, and either the first witness or the third witness lied. The first witness did not lie. Therefore the second and third witnesses both lied. (F, S, T)

5. Either Socrates is mortal, or he's not immortal. Therefore, Socrates is mortal. (M)

6. Either our sense of sight deceives us, or else our senses of sight and hearing both deceive us. Therefore our sense of sight deceives us. (S, H)

7. Either time had a beginning, or time is not finite. Either time will end, or time is not finite. Therefore either time is not finite, or else time had a beginning and will end. (B, F, E)

8. Either three times one-third is three times .333···, or else one-third is not .333···. One-third *is* .333···, and either three times one-third is different from three times .333··· or else one is .999···. Therefore, one is not different from .999···. (A, B, C)

9. You should not open the bidding with "Two no-trump" unless you have 22–24 high-card points and a stopper in every suit. You have 22–24 high-card points but you do *not* have a stopper in every suit. Therefore even though you have 22–24 high-card points, you should not open the bidding with "Two no-trump." $(B, H, S$; recall the discussion of 'unless' in Section 1.4)

10. The valence of the calcium ion is either +2 or +3, and the valence of the phosphate ion is either −2 or −3. So either the valence of the calcium ion is +3, or the valence of the phosphate ion is −2, or else the valence of the calcium ion is +2 and the valence of the phosphate ion is −3. (C, D, P, Q)

11. Either silver stocks are going up or silver bullion prices are not; and either the Dow Jones Transportation Index is going up or the Dow Jones Industrial Index is not. Silver bullion prices are going up and so is the Dow Jones Industrial Index. Therefore silver stocks are going up and so is the Dow Jones Transportation Index. (S, B, T, I)

12. Francis Bacon wrote *Hamlet*. So either Bacon wrote *King Lear*, or Shakespeare wrote *King Lear*, or else neither Bacon nor Shakespeare wrote *King Lear*. (H, B, S)

13. The set of prime numbers is either denumerable or countable. The set of prime numbers is either finite, denumerable or uncountable. Therefore the set of prime numbers is either denumerable or finite. (D, C, F)

14. Either Neanderthal man was especially well adapted to living in

cold climates, or else he did not live in Europe during the Würm glaciation. Unless the carbon-14 studies of his remains are inaccurate, Neanderthal man did live in Europe during the Würm glaciation. Therefore either Neanderthal man was especially well adapted to living in cold climates, or the carbon-14 studies of his remains are inaccurate. (A, L, C)

*15. Either no evil exists in the world, or God is unwilling to prevent its existence, or He is unable to prevent its existence. Either God is both willing and able to prevent the existence of evil, or else He is neither willing nor able to prevent its existence. Therefore either no evil exists in the world or else God is both unwilling and unable to prevent its existence. (E, W, A)

Deductive Validity and Theorems

We have confined our attention in this section to arguments whose validity hinges solely upon the behavior of '&', 'v', and '~'. In connection with such arguments it is useful to introduce the special terms 'deductively valid' and 'deductively invalid':

DEFINITION An argument is **deductively valid** if and only if its conclusion can be deduced from its premises by repeated use of the rules **Conjunction, Simplification, Addition, Disjunctive Syllogism, Excluded-Middle Introduction, Association, Distribution,** Commutation, **Double Negation,** and **De Morgan**; otherwise, the argument is **deductively invalid**.

Thus the argument 'The universe is not both non-empty and infinite; it is non-empty; therefore the universe is finite' is deductively valid, as we can show by deducing its conclusion from its premises:

1. $\sim(\sim E \, \& \sim F)$ Premise
2. $\sim E$ Premise   DEDUCE: F
3. $\sim\sim E$ v $\sim\sim F$ 1 De M
4. E v $\sim\sim F$ 3 DN
5. $\sim\sim F$ 2, 4 DS
6. F 5 DN

Sometimes it is convenient to employ a sort of shorthand in discussing deductive validity. We write, for example,

$\sim(\sim E \, \& \sim F), \sim E \vdash F$

instead of the wordier

The argument with premises '$\sim(\sim E$ & $\sim F)$' and '$\sim E$' and conclusion 'F' is deductively valid,

that is, instead of writing

'F' is deducible (using our ten rules of inference and replacement) from '$\sim(\sim E$ & $\sim F)$' and '$\sim E$'.

It should always be kept in mind, of course, that in deducing the conclusion of an argument we are *not* establishing that it is *true* but only that it would have to be true *if* the *premises* were true.

We have already seen that our rules can be used to deduce the conclusions of many arguments from their premises. But sometimes they can be used to do something else. Certain sentences can be deduced, using just our rules, *from no premises at all*. What happens in the case of such a sentence is that we use **E-M I** at the beginning of the deduction, and then go on to deduce the sentence in question from the one **E-M I** starts us off with, for example:

1. $F \vee \sim F$	E-M I	
2. $(F \vee \sim F) \vee E$	1 Add	
3. $E \vee (F \vee \sim F)$	2 Com	
4. $(E \vee F) \vee \sim F$	3 Assoc	
5. $(F \vee E) \vee \sim F$	4 Com	

What has been shown here is very different from what we show when the conclusion of an argument is deduced from its premises. The latter sort of deduction establishes only that a certain sentence must be true *if* certain others are. But here no premises at all are involved; instead, the truth of 'Either the universe is finite, or it is empty, or it is not finite' is established *unconditionally*. The sentence in question has, in effect, been shown to be a *logical truth*, that is, true on logical grounds alone.

We shall call expressions which can be obtained in this way — as last lines of deductions in which *no* premises are involved — the **theorems** of our deductive system. Here, again, it is convenient to employ shorthand. Just as we write

$E \vee F, \sim E \vdash F$

as an abbreviation for

'*F*' is deducible (using the rules of this section) from the premises '*E* v *F*' and '~*E*'

so will we write, for example,

$\vdash (F$ v $E)$ v $\sim F$

as an abbreviation for

'(*F* v *E*) v ~*F*' is deducible (using the rules of this section) from no premises at all,

that is, for

'(*F* v *E*) v ~*F*' is a theorem (of our deductive system).

Deductions in which *no* premises are involved, every line being justified instead by one of our rules of inference or replacement, are often referred to as **proofs**. Every theorem, then, is the last line of some proof. Naturally we can expect the proofs to get more complicated as the theorems become more intricate:

1. *B* v ~*B*	E-M I
2. *A* v ~*A*	E-M I
3. (*B* v ~*B*) & (*A* v ~*A*)	1, 2 Conj
4. [(*B* v ~*B*) & *A*] v [(*B* v ~*B*) & ~*A*]	3 Dist
5. [*A* & (*B* v ~*B*)] v [(*B* v ~*B*) & ~*A*]	4 Com
6. [*A* & (*B* v ~*B*)] v [~*A* & (*B* v ~*B*)]	5 Com
7. [(*A* & *B*) v (*A* & ~*B*)] v [~*A* & (*B* v ~*B*)]	6 Dist
8. [(*A* & *B*) v (*A* & ~*B*)] v [(~*A* & *B*) v (~*A* & ~*B*)]	7 Dist

Theorems, and their proofs, are not only interesting in their own right; they will play an important role in Chapter 4 as well.[4]

---

[4]The reader may wish to compare the sign '$\vdash$' (and the concepts of *deductive validity* and *theorem*) with the sign '$\models$' (and the concepts of *truth-table validity* and *tautology*) introduced in Sections 2.1 and 2.3.

EXERCISES

I.   Show that each of the following is a theorem of the deductive system developed in this section:
1. $\sim F \lor F$ 3. $\sim F \lor (\sim E \lor F)$
2. $\sim (F \,\&\, \sim F)$ 4. $(E \,\&\, F) \lor (\sim E \lor \sim F)$

II.   Answer the following questions, *explaining* each of your answers:
1.   If $\mathscr{P}$ and $\mathscr{Q}$ are expressions which are theorems of our system, must the corresponding expression

$$\mathscr{P} \,\&\, \mathscr{Q}$$

also be a theorem?
2.   If $\vdash \mathscr{P}$ and

$$\vdash \sim (\mathscr{P} \,\&\, \mathscr{Q})$$

must $\mathscr{Q}$ be a theorem as well?
3.   If $\mathscr{P}$ and $\mathscr{Q}$ are expressions such that

$$\vdash \sim \mathscr{P} \lor \mathscr{Q}$$

must it be the case that

$$\mathscr{P} \vdash \mathscr{Q}?$$

3.2 Rules of Inference and Replacement for '⊃' and '≡'

Many of the arguments we have looked at thus far seem to have been worded in an especially contrived manner, for example:

Either time had a beginning or time is not finite.	$B \lor \sim F$
Either time will end or time is not finite.	$E \lor \sim F$
Therefore either time is not finite or else time had a beginning and will end.	$\therefore \sim F \lor (B \mathbin{\&} E)$

It would be far more natural to reword both the premises and the conclusion, and present the argument this way:

If time is finite then time had a beginning.	$F \supset B$
If time is finite then time will end.	$F \supset E$
Therefore if time is finite then time had a beginning and will end.	$\therefore F \supset (B \mathbin{\&} E)$

Of course we want to extend our treatment of '&', 'v', and '~' so that we can handle arguments presented in this sort of language, too, and that requires us to find and add to our set of ten rules for '&', 'v', and '~' some rules of inference and replacement for '⊃' and '≡'. We begin by noticing that each of the following arguments is valid in the sense — discussed in Chapter 1 — that their conclusions must be true if their premises are.

If I think then I exist.
I think.
Therefore I exist.

If all killing is wrong then killing in defense of one's children is wrong.
It is not the case that killing in defense of one's children is wrong.
Therefore it is not the case that all killing is wrong.

If God is omniscient then God always knows what time it is.
If God always knows what time it is then what God knows at any given moment differs from what He knows at any other moment.
Therefore if God is omniscient then what God knows at any given moment differs from what He knows at any other moment.

Either time is infinite or time is circular.
If time is infinite then time had no beginning.
If time is circular then time has no end.
Therefore either time had no beginning or time has no end.

These arguments are, respectively, of the forms:

$p \supset q$	$p \supset q$	$p \supset q$	$p \vee q$
p	$\sim q$	$q \supset r$	$p \supset r$
$\therefore q$	$\therefore \sim p$	$\therefore p \supset r$	$q \supset s$
			$\therefore r \vee s$

Clearly, no argument of any of these forms could have true premises but a false conclusion. Accordingly, we add to the five rules of inference that we already have (**Conjunction, Simplification, Addition, Disjunctive Syllogism,** and **Excluded-Middle Introduction**) the following four:

Rules of Inference	*Justification*[5]

Modus Ponens (MP)

In any deduction, one may infer from two previously obtained sentences of the forms

$$p \supset q$$

and

$$p$$

the corresponding sentence

$$q.$$

Any argument of the form

$$p \supset q$$
$$p$$
$$\therefore q$$

is valid, for if a conditional is true and its antecedent is also, then its consequent must be.

Modus Tollens (MT)

In any deduction, one may infer from two previously obtained sentences of the forms

$$p \supset q$$

and

$$\sim q$$

the corresponding sentence of the form

$$\sim p.$$

Any argument of the form

$$p \supset q$$
$$\sim q$$
$$\therefore \sim p$$

is valid, for if a conditional is true but its consequent is not then its antecedent cannot be.

---

[5]These justifications may be spelled out in more detail by referring explicitly to the definition of a material conditional given in Section 1.4 (see Exercise IV below). They are elaborated in a different way, by truth-tables, in Section 2.3.

Hypothetical Syllogism (HS)

In any deduction, one may infer from two previously obtained sentences of the forms

$$p \supset q$$

and

$$q \supset r$$

the corresponding sentence of the form

$$p \supset r.$$

Any argument of the form

$$p \supset q$$
$$q \supset r$$
$$\therefore p \supset r$$

is valid — if the first two conditionals are true then the third one must also be true.

Constructive Dilemma (CD)

In any deduction, one may infer from three previously obtained sentences of the forms

$$p \lor q$$
$$p \supset r$$

and

$$q \supset s$$

the corresponding sentence of the form

$$r \lor s.$$

Any argument of the form

$$p \lor q$$
$$p \supset r$$
$$q \supset s$$
$$\therefore r \lor s$$

is valid, for if one or the other of two true conditionals has a true antecedent then one or the other of them has a true consequent.

Moreover, it should be clear that *neither* of the following forms of argument is valid and hence they should never be employed in deductions:

$$p \supset q \qquad p \supset q$$
$$q \qquad\qquad \sim p$$
$$\therefore p \qquad\qquad \therefore \sim q$$

The first is sometimes confused with **Modus Ponens** and the second with **Modus Tollens**. If we replace '*p*' with the false sentence 'More than twelve planets orbit the sun' and '*q*' with the true sentence 'More than eight planets orbit the sun', however, both resulting arguments have true premises but false conclusions:

If more than twelve planets orbit the sun, then more than eight planets orbit the sun.

If more than twelve planets orbit the sun, then more than eight planets orbit the sun.

More than eight planets orbit the sun.
Therefore more than twelve planets orbit the sun.

Not more than twelve planets orbit the sun.
Therefore not more than eight planets orbit the sun.

We will need some additional rules of replacement, as well, and the considerations of Section 1.4 suggest the following:

Transposition (Trans)

$p \supset q :: {\sim}q \supset {\sim}p$

Implication (Impl)

$p \supset q :: {\sim}p \vee q$

Exportation (Exp)

$(p \;\&\; q) \supset r :: p \supset (q \supset r)$

Equivalence (Equiv)

$p \equiv q :: (p \supset q) \;\&\; (q \supset p)$
$p \equiv q :: (p \;\&\; q) \vee ({\sim}p \;\&\; {\sim}q).$[6]

More thoroughgoing justifications for these rules are provided by the methods of Section 2.2.

With these additional rules we can now handle such arguments as:

1. $C \supset ({\sim}P \supset A)$ — Premise — If *abstract* entities are those that cannot be perceived with the senses, then if the number 89 cannot be perceived with the senses than the number 89 is an abstract entity.

2. $N \supset (A \supset E)$ — Premise — If the number 89 exists, then if it is an abstract entity then of course at least one abstract entity exists.

3. ${\sim}N \supset {\sim}B$ — Premise — If the number 89 does not exist, then there are no prime numbers between 84 and 96.

4. B — Premise — There is a prime number between 84 and 96.

$\therefore C \supset (P \vee E)$ — Therefore if abstract entities are those that cannot be perceived with the senses then either the number 89 can be perceived with the senses or else at least one abstract entity exists.

---

[6]The 'if ... then ---' locution is undoubtedly used, like most locutions, in a variety of different senses in ordinary English discourse. With a single set of rules like ours, of course, we get at only one of the ways in which it can be used. Presumably *other* sets of rules could be constructed that would capture *other* senses of the locution in question. The reader interested in attempts to construct other such sets of rules might begin with the references cited in Section 2.2, footnote 5.

We first use **MT** to play (3) and (4) off against one another

5. ~~B	4 DN	
6. ~~N	3, 5 MT	

and notice that **DN** will permit us to use **MP**:

7. N	6 DN	
8. A ⊃ E	7, 2 MP	

If the consequent of (1) were the same as the antecedent of (8) we could use **HS**; of course this is not the case, but if we use **Exportation** on (1) first, we'll be able to use **HS** anyway:

9. (C & ~P) ⊃ A	1 Exp	
10. (C & ~P) ⊃ E	9, 8 HS	

We could now use **Implication** on (10) to get '~(C & ~P) v E'; **De Morgan** would then lead us to '(~C v ~~P) v E', **DN** and **Assoc** would give us '~C v (P v E)' and **Implication** would finish the job. But we can save a couple of steps by using **Exportation** on (10) first:

11. C ⊃ (~P ⊃ E)	10 Exp	
12. C ⊃ (~~P v E)	11 Impl	
13. C ⊃ (P v E)	12 DN	

It is generally good strategy to use **MP**, **MT**, **HS**, and **CD** as often as possible, often going out of one's way via **DN**, **Implication**, **Exportation**, **Transposition**, and the like to develop opportunities to apply such rules. Presented with

1. A ⊃ (B & C)	Premise	
2. D ⊃ E	Premise	
3. A v D	Premise	
4. B ⊃ F	Premise	
5. (~E ⊃ F) ⊃ (~G ⊃ ~H)	Premise	DEDUCE: H ⊃ G

it is probably best to notice first that (3) is the disjunction of the ante-cedents of (1) and (2):

6. $(B \text{ \& } C) \vee E$ 1, 2, 3 CD

The conclusion we're after, '$H \supset G$', is concealed in the consequent of (5), and one might notice this next:

7. $(\sim E \supset F) \supset (H \supset G)$ 5 Trans

So we need '$\sim E \supset F$'. Since no obvious way of obtaining it suggests itself, we look around for something else to do. If we commute (6) we'll be able to use **Dist**, of course, and experience in the preceding section has surely taught us to give this route a chance:

8. $E \vee (B \text{ \& } C)$ 6 Com

9. $(E \vee B) \text{ \& } (E \vee C)$ 8 Dist

Now which half of (9) is most likely to get '$\sim E \supset F$' for us? Let's see what we can do with

10. $E \vee B$ 9 Simp

first—if it leads nowhere we can always turn to '$E \vee C$' later. (10) is a disjunction but there's no way to use it in a **Constructive Dilemma**; so instead of leaving it as it is let's use **Implication** to turn it into a con-ditional:

11. $\sim\sim E \vee B$ 10 DN

12. $\sim E \supset B$ 11 Impl

That was the right road to take! (12) has the same consequent as (4) has antecedent, so **HS** and **MP** finish our deduction:

13. $\sim E \supset F$ 12, 4 HS

14. $H \supset G$ 13, 7 MP

The deductions we've looked at so far involve no biconditionals. Those that do generally require one or more uses of our rule **Equivalence**:

1. ~E v [(W & A) v (~W & ~A)]   Premise
2. E   Premise   DEDUCE: A ≡ W
3. E ⊃ [(W & A) v (~W & ~A)]   1 Impl
4. E ⊃ (W ≡ A)   3 Equiv
5. W ≡ A   2, 4 MP
6. (W ⊃ A) & (A ⊃ W)   5 Equiv
7. (A ⊃ W) & (W ⊃ A)   6 Com
8. A ≡ W   7 Equiv

It will later be convenient to add to our system one last rule, the rule of **Conditional Proof**, which will correspond to a very natural way of arguing and will have the advantage of shortening many deductions considerably. But it will turn out (see Chapter 4) that we *need* not add that rule; we shall be able to prove nothing with it that we could not have established using only the rules already presented in this chapter.

EXERCISES

I.   Finish annotating the following two deductions:

    1.   1.   (A & B) ⊃ C   Premise
         2.   A   Premise
         3.   C ⊃ (D & E)   Premise   DEDUCE:
         4.   A ⊃ (B ⊃ C)   ~B v (D & E)
         5.   B ⊃ C
         6.   B ⊃ (D & E)
         7.   ~B v (D & E)

    2.   1.   ~(J & K)   Premise
         2.   ~(H v ~I) ⊃ (J & K)   Premise
         3.   H ⊃ ~F   Premise
         4.   ~I ⊃ ~G   Premise
         5.   ~F ⊃ G   Premise   DEDUCE:
         6.   ~~(H v ~I)   F ≡ ~G
         7.   H v ~I
         8.   ~F v ~G
         9.   F ⊃ ~G
        10.   ~G ⊃ ~~F
        11.   ~G ⊃ F
        12.   (F ⊃ ~G) & (~G ⊃ F)
        13.   F ≡ ~G

II. Deduce the conclusions of each of the following arguments from their premises:

1. $(D \supset C) \& (A \supset B)$
   $C \& A$
   $\therefore B \vee D$

2. $\sim(A \& \sim R)$
   $\sim(R \& S)$
   $\therefore S \supset \sim A$

3. $(R \supset S) \vee T$
   $\sim(S \vee T)$
   $\therefore \sim R \vee \sim Q$

4. $L \supset N$
   $(\sim N \supset \sim L) \supset (M \supset P)$
   $\sim P$
   $\therefore \sim M$

5. A
   $\sim B$
   $\therefore B \equiv \sim A$

6. $S \supset (M \supset W)$
   $P \supset (M \& S)$
   $\therefore \sim P \vee W$

7. $N \vee \sim E$
   $\sim A \vee E$
   $\therefore N \vee \sim A$

8. $\sim E \vee (\sim W \vee \sim A)$
   $(W \& A) \vee (\sim W \& \sim A)$
   $\therefore \sim E \vee (\sim W \& \sim A)$

9. $(Z \& M) \supset (S \vee A)$
   $Z \supset \sim S$
   $\therefore (Z \& \sim A) \supset \sim M$

10. $K \equiv Q$
    $(K \& V) \equiv \sim(Q \vee V)$
    $\therefore K \supset \sim V$

III. Symbolize each of the following arguments; then deduce their conclusions from their premises:

1. The number 91 is prime just in case it is not composite. The number 91 is not both prime and divisible by 7. The number 91 is divisible by 7. Therefore the number 91 is composite. (P, C, D)

2. I'm damned if I do, and damned if I don't. Therefore I'm damned. (M, D)

3. If argument (1) has a false premise, it is unsound. If argument (1) is invalid, it is unsound. So if argument (1) either has a false premise or is invalid, then it is unsound. (F, S, V)

4. If argument (1) either has a false premise or is invalid, it is unsound. Consequently, if argument (1) has a false premise it is unsound and if argument (1) is invalid it is unsound. (F, V, S)

5. If the President sounds optimistic then the stock market will go up. So if the economic news is bad but the President sounds optimistic, the stock market will go up. (O, U, B)

6. If argument (5) is sound, then it is valid and its premise is true. Thus, if argument (5) is sound then it is valid, and if argument (5) is sound then its premise is true. (S, V, P)

7. If argument (5) is sound then it is valid. And if argument (5) is sound then its premise is true. Therefore if argument (5) is sound then it is valid and its premise is true. (S, V, P)

8. If my partner's lead of the queen of hearts was not a mistake, then he has either the heart king or the heart jack as well. If he

has the heart king, then he made a mistake when he failed to respond to my opening bid. So if my partner made no mistake when he failed to respond to my opening bid and made no mistake either in leading the queen of hearts, he has the heart jack. (*Q*, *K*, *J*, *B*)

9. If God can create a stone too heavy for Him to lift, then there is something God cannot do. If it is not the case that God can create a stone too heavy for Him to lift, then there is something God cannot create. If there is something God cannot do then God is not omnipotent, and if there is something He cannot create then He is not omnipotent. Therefore God is not omnipotent. (*S*, *D*, *C*, *O*)

10. If my plans are going to work out, then I'm going to get a D on the first test but go into the final with a high average. If I study fairly hard but skip the hard problems, then I'll get a B on the second test. If I get a B on the second test but a D on the first, I won't go into the final with a high average. Therefore if I study fairly hard, then if I skip the hard problems my plans are not going to work out. (*P*, *D*, *H*, *F*, *S*, *B*)

11. If the price of the stock rises we'll make money on the shares we own, and if the price of the stock falls we'll make money on warrants we sold short. So if the price of the stock either rises or falls, we'll make money on either the shares we own or the warrants we sold short. (*R*, *S*, *F*, *W*)

12. The set of prime numbers is either denumerable or countable. If the set of prime numbers is countable, then it is either finite or denumerable. The set of prime numbers is not finite. It follows, then, that the set of prime numbers is denumerable. (*D*, *C*, *F*)

IV. Show in detail how each of the rules of inference and replacement introduced in this section may be justified by the characterizations of *material conditionals* and *material biconditionals* provided in Section 1.4.

3.3 Conditional Proof[7]

Consider the argument

If the set of prime numbers is finite, then the set of prime numbers has a smallest member and a largest member.	$F \supset (S \;\&\; L)$
Therefore if the set of prime numbers is finite then it has a largest member.	$\therefore F \supset L$

And imagine how one might think this argument through to himself, if he were being careful to try to make each step explicit:

I'm told that if the set of prime numbers is finite then it has both a smallest and a largest member, and I'm asked to deduce from this that if the set of prime numbers is finite then it has a largest member. What if I assume for a moment that the set of prime numbers is finite; could I then show that it has a largest member? Why of course, for there is practically nothing to show. It follows from my assumption and the premise of the original argument, by **MP**, that the set of primes has both a smallest and a largest member; and from this, in turn, it follows by **Com** and **Simp** that it has a largest member, which is what I had to show! Summing up: by assuming the antecedent of the conclusion, 'The set of prime numbers is finite', I was able to deduce its consequent, 'The set of prime numbers has a largest member'. So I *have* established that *if* the set of primes is finite then it has a largest member.

Let's find a way of writing up this argument clearly. We can begin, as usual, by writing the premise as line (1) of our deduction, noting the conclusion off to the right:

1. $F \supset (S \;\&\; L)$ Premise   DEDUCE: $F \supset L$

The first step in our argument was to assume (temporarily) that the antecedent of the conclusion is true. This is, of course, not something we can *deduce* from (1). To make it obvious that this is merely an *assumption* which we're making for the moment, we'll label it **Assumption**, indent it to the right and point an arrow at it from the left:

1. $F \supset (S \;\&\; L)$ Premise   DEDUCE: $F \supset L$

$\mapsto$ 2. F Assumption

---

[7]We are indebted to one of the publisher's anonymous reviewers for the essentials of the formulation of the rule of conditional proof given at the end of this section and to Robert Neidorf, *Deductive Forms*, New York: Harper and Row, 1967, for teaching us how best to introduce and motivate this rule and its typographical trappings.

From the premise, (1), and our assumption, (2), we obtain

3. *S* & *L*   1, 2 MP

To make it clear that (3) has only been deduced with the help of our assumption, (2), we indent line (3) also, and allow the tail of the arrow marking that assumption to run down beside it:

1. *F* ⊃ (*S* & *L*)   Premise   DEDUCE: *F* ⊃ *L*

2. *F*   Assumption

3. *S* & *L*   1, 2 MP

So long as we continue to make deductions in the presence of our assumption rather than from our premise alone, we indent the lines we deduce and continue the tail of the arrow alongside:

1. *F* ⊃ (*S* & *L*)   Premise   DEDUCE: *F* ⊃ *L*

2. *F*   Assumption

3. *S* & *L*   1, 2 MP

4. *L* & *S*   3 Com

5. *L*   4 Simp

Given (1), we have now shown that *if* (2) is true *then* (5) must be true as well. We have established the conditional '*F* ⊃ *L*', by actually deducing its consequent from its antecedent (in the presence of our premise). To show that '*F* ⊃ *L*' follows from (1) alone, and that our assumption (2) has now served its purpose and can be abandoned, we *close off* the assumption by extending the tail of the arrow horizontally under (5) and writing '*F* ⊃ *L*' beneath—thus indicating that the assumption made at line (2) is no longer operative. We call the rule permitting this move the **Rule of Conditional Proof (CP)** and delay its precise formulation until the end of this section.

1. *F* ⊃ (*S* & *L*)   Premise   DEDUCE: *F* ⊃ *L*

2. *F*   Assumption

3. *S* & *L*   1, 2 MP

4. *L* & *S*   3 Com

5. *L*   4 Simp

6. *F* ⊃ *L*   2–5 CP

Notice that the justification for line (6) refers to all steps in the subordinate deduction running from line (2) *through* line (5). It is because of this entire *sequence* that we are sure (6) follows from (1), for the lines of that sequence show that given our premise, the consequent of (6) must be true if its antecedent is.

Of course we could have deduced (6) from (1) using just the rules we already had. The deduction would not have been any longer (though in most other cases it would be):

1. $F \supset (S \& L)$	Premise	DEDUCE: $F \supset L$
2. $\sim F \lor (S \& L)$	1 Impl	
3. $(\sim F \lor S) \& (\sim F \lor L)$	2 Dist	
4. $(\sim F \lor L) \& (\sim F \lor S)$	3 Com	
5. $\sim F \lor L$	4 Simp	
6. $F \supset L$	5 Impl	

However, this deduction is more difficult to discover than the former. It also seems considerably less natural and more contorted, for the first deduction follows much more closely the steps one might actually go through in quickly thinking his way from premises to conclusion.

EXERCISE

Use **CP** to construct new deductions for (9) and (10) in group (II) and for (3), (5), (8), and (11) in group (III) of the Exercises at the end of Section 3.2.

CP is more versatile than we have shown so far. For one thing, an assumption for **CP** need not be introduced immediately after the premises. Furthermore, the conditional established by **CP** need not be the conclusion of the original argument.

1. $\sim A \lor B$	Premise	
2. $\sim A \lor (\sim B \lor C)$	Premise	DEDUCE: $\sim A \lor C$
3. $A \supset B$	1 Impl	
4. $A \supset (\sim B \lor C)$	2 Impl	
5. $A \supset (B \supset C)$	4 Impl	

6. *A*	Assumption	
7. *B*	3, 6 MP	
8. *B* ⊃ *C*	5, 6 MP	
9. *C*	7, 8 MP	
10. *A* ⊃ *C*	6–9 CP	
11. ~*A* v *C*	10 Impl	

One is permitted to assume any sentence at any point in a deduction, provided only that each assumption made is later closed off. At line (6) here, for example, we *need* not have assumed '*A*'—we could have assumed '~*A*' or '*C*' or even '*M*'. But there was no reason to make any of these other assumptions. Had we assumed '*M*', say, then when we later gave up that assumption we would be left with an expression beginning '*M* ⊃ ...', and no such expression would have helped in obtaining '~*A* v *C*'. We chose instead to assume '*A*' and work for '*C*', because we knew that **CP** would then give us '*A* ⊃ *C*' and that '~*A* v *C*' would follow immediately.

It is also useful to be able to stack several conditional proofs on top of one another, using **CP** once to establish one thing we need and then using it again to establish something else, as in:

1. *A* ⊃ *B*	Premise	
2. (*A* & *B*) ⊃ *C*	Premise	DEDUCE: [*A* ⊃ (*A* & *C*)] & [(*A* & *C*) ⊃ *A*]
3. *A*	Assumption	
4. *B*	1, 3 MP	
5. *A* & *B*	3, 4 Conj	
6. *C*	2, 5 MP	
7. *A* & *C*	3, 6 Conj	
8. *A* ⊃ (*A* & *C*)	3–7 CP	
9. *A* & *C*	Assumption	
10. *A*	9 Simp	
11. (*A* & *C*) ⊃ *A*	9–10 CP	
12. [*A* ⊃ (*A* & *C*)] & [(*A* & *C*) ⊃ *A*]	8, 11 Conj	

Double use of **CP** is especially common when the conclusion of the argument one is concerned with happens to be a biconditional.

1. $A \supset B$	Premise	
2. $B \supset C$	Premise	DEDUCE: $A \equiv (A \& C)$
→3. A	Assumption	
4. B	1, 3 MP	
5. C	2, 4 MP	
6. $A \& C$	3, 5 Conj	
7. $A \supset (A \& C)$	3–6 CP	
→8. $A \& C$	Assumption	
9. A	8 Simp	
10. $(A \& C) \supset A$	8–9 CP	
11. $[A \supset (A \& C)]$ & $[(A \& C) \supset A]$	7, 10 Conj	
12. $A \equiv (A \& C)$	11 Equiv	

Line (8) here does entitle us to line (9), of course, but notice that it would have been a *mistake* to have tried to justify (9) on the basis of (6). (6) was only obtained while the assumption made at line (3) was in force and that assumption is no longer operative at line (9), having been given up earlier. In general, once an assumption has been closed off no later lines of the deduction may be justified by reference to that assumption nor to any of the lines occurring between it and the point at which it was closed off. The little three-sided boxes formed by the arrows and their tails serve to keep track of such matters.

There are times, too, when we will want to nest conditional proofs inside one another, as in:

1. $A \supset (B \supset C)$	Premise	
2. $B \supset (C \supset D)$	Premise	DEDUCE: $A \supset (B \supset D)$
→3. A	Assumption	
4. $B \supset C$	1, 3 MP	
→5. B	Assumption	
6. C	4, 5 MP	
7. $C \supset D$	2, 5 MP	
8. D	6, 7 MP	
9. $B \supset D$	5–8 CP	
10. $A \supset (B \supset D)$	3–9 CP	

This sort of thing happens most often when **CP** is used, as it frequently is, to establish *theorems*, for example:

→ 1. $A \supset B$		Assumption
→ 2. $B \supset C$		Assumption
3. $A \supset C$		1, 2 HS
4. $(B \supset C) \supset (A \supset C)$		2–3 CP
5. $(A \supset B) \supset [(B \supset C) \supset (A \supset C)]$		1–4 CP

In nesting conditional proofs this way, one must be sure that the order in which assumptions are closed off is the reverse of that in which they were introduced. This restriction, as well as the others mentioned above, are packed into the following more formal statement of the rules developed in this section, which serves also as a summary.

Rule of Assumptions (Assumption)

Any sentence may be introduced as a new step in a deduction, *provided that*:
   i.   it is indented to the right, and a new vertical line is introduced on the left at that point, with an arrow at its top pointing to the sentence being introduced;
   ii. the vertical line in question is eventually discontinued by an application of the rule of **CP** (see below), but continued up to that point.

Rule of Conditional Proof (CP)

Let 'q' be a step in a deduction with a continuous vertical line passing *immediately* to its left, and let 'p' be the sentence heading that vertical line. Then the corresponding sentence of the form '$p \supset q$' may be introduced as the next step, *provided that*:
   i.   the vertical line in question is discontinued by extending it horizontally to the right immediately below 'q' and above '$p \supset q$';
   ii. all steps beside which this vertical line passes are thereby regarded as sealed off from the rest of the deduction (that is, no *later* steps in the deduction are justified by a reference to any of these earlier ones enclosed now by a three-sided box).

EXERCISES

   I.   Use **CP** to construct new deductions for (6), (7), and (8) in group (II) and for (4), (6), and (10) in group (III) of the Exercises at the end of Section 3.2.

II. Construct deductions for the following, using **CP** when helpful:

1. $(E \& W) \equiv (E \& A)$
   $\therefore E \supset [(W \& A) \vee (\sim W \& \sim A)]$

2. $(D \vee P) \supset (G \& F)$
   $(T \supset G) \supset [(R \vee B) \supset E]$
   $(\sim S \supset H) \supset [B \& \sim(E \& U)]$
   $\therefore D \supset (S \supset \sim U)$

3. Jones has not learned how to use correctly the expression "Smith acted of his own free will" unless he has seen Smith acting freely and heard others use the expression "Smith is acting of his own free will." If all of our actions are determined according to physical laws, then no one ever acts of his own free will and Jones has never seen Smith acting freely. So if Jones has learned how to use correctly the expression "Smith acted of his own free will," not all of our actions are determined according to physical laws. (L, S, H, D, N)

4. Al's claim is true if and only if Bill's is. Therefore Bill's claim is untrue if and only if Al's is. (A, B)

5. If either Al's claim or Bill's claim is true, then Al's claim is true if and only if Bill's claim is true. It follows that Al's claim is true if and only if Bill's is. (A, B)

6. If the Administration pursues a tight-money policy, then either the leaders of the opposition party support a tight-money policy or else they denounce governmental meddling. If the leaders of the opposition party support a tight-money policy and the economy goes steadily downhill, then they'll denounce governmental meddling and claim that their party could get the economy back on its feet. So if the Administration pursues a tight-money policy or the leaders of the opposition party support a tight-money policy, then if the economy goes steadily downhill the leaders of the opposition party will denounce governmental meddling. (P, S, D, H, C)

7. If Yossarian flies his missions, then if knowingly placing one's life in danger is irrational, then Yossarian is irrational. If Yossarian asks to be grounded and a request to avoid a perilous situation is rational, then Yossarian is rational. A request to avoid a perilous situation is rational if and only if knowingly placing one's life in danger is irrational. Therefore, if Yossarian flies his missions but asks to be grounded, then knowingly placing one's life in danger is rational and a request to avoid a perilous situation is irrational. (F, K, R, G, A)

III. Use **CP** to establish each of the following theorems:

1. $[(A \supset A) \supset B] \supset B$

2. $[(A \supset B) \supset C] \supset$
   $[(C \supset A) \supset (D \supset A)]$

3. $[(A \supset B) \supset A] \supset A$

4. $(A \supset B) \vee (B \supset C)$

3.4 Indirect Proof

Consider the argument

Either the set of even numbers is infinite, or else the set of even numbers and the set of odd numbers are both infinite.	$E \vee (E \ \& \ O)$
If the set of even numbers is infinite then for each even number there exists a larger even number.	$E \supset L$
Therefore for each even number there exists a larger even number.	$\therefore L$

One natural way to argue, in English, from those premises to that conclusion would go something like this:

Given the two premises, 'L' must be true. For suppose not. Then it would follow, with premise two, that 'E' is not true, either. But then it follows from premise one, since 'E' is false, that 'E' and 'O' are both true and so, in particular, that 'E' is. Thus, if 'L' were *not* true, 'E' would be both true and false. Since 'E' cannot be both true and false, our original supposition — that 'L' is *not* true — must have been false, and we can conclude that 'L' is true.

The reader may be familiar with this way of arguing from a high-school course in geometry or readings in number theory, where one commonly establishes the truth of a conclusion by supposing it false and showing that such a supposition leads to an explicit contradiction, that is, a sentence of the form 'p & $\sim p$'. He may be surprised to find that the deduction informally described above translates quite directly into one of our more formal deductions:

Given the two premises,	1. $E \vee (E \ \& \ O)$	Premise	
'L' must be true.	2. $E \supset L$	Premise	DEDUCE: L
For suppose not.	→ 3. $\sim L$	Assumption	
Then it would follow, with premise two, that 'E' is not true, either.	4. $\sim E$	2, 3 MT	
But then it follows from premise one, since 'E' is false, that 'E' and 'O' are both true.	5. $E \ \& \ O$	1, 4 DS	
And so, in particular, that 'E' is.	6. E	5 Simp	
	7. $E \ \& \ \sim E$	6, 4 Conj	
Thus, if 'L' were *not* true, 'E' would be both true and false.	8. $\sim L \supset (E \ \& \ \sim E)$	3–7 CP	

Since '*E*' cannot be both	9. ~*E* v ~~*E*	E-M I
true and false,	10. ~(*E* & ~*E*)	9 De M
our original supposition—that '*L*' is not true—must have been false,	11. ~~*L*	8, 10 MT
and we can conclude that '*L*' is true.	12. *L*	11 DN

Notice that the overall strategy is quite natural. To establish that our conclusion must be true if the premises are, we show that its negation is *inconsistent* with them, that is, that it is *impossible* for the premises to be true while the negation of the conclusion is true also. Thus we show the argument to be valid.

Some authors short-cut the above deduction by introducing a special rule, the *Rule of Indirect Proof* (or the *Rule of Reductio ad Absurdum*). Roughly, it permits one to end deductions like the one just given immediately after the contradiction has been obtained. We will *not* introduce such a special rule, preferring to regard ourselves as having merely discovered a new *strategy* useful in constructing deductions.

Even where other strategies can also be employed, indirect proof is often the simplest to use. Given the argument

1. *D* v *C*	Premise	
2. *C* ⊃ (*F* v *D*)	Premise	
3. ~*F*	Premise	DEDUCE: *D*

we can deduce the conclusion "indirectly" by beginning with the assumption of its negation:

4. ~*D*	Assumption	

DS is suggested immediately,

5. *C*	4, 1 DS	

and it sets up

6. *F* v *D*	5, 2 MP	

One more application of **DS** gives us our contradiction,

7. *D*	3, 6 DS	
8. *D* & ~*D*	7, 4 Conj	

and once the contradiction has been obtained we finish, as always when using indirect proof, with the sequence: **CP**, **E-M I**, **De M**, **MT**.

9. $\sim D \supset (D \,\&\, \sim D)$ 4–8 CP

10. $\sim D \vee \sim\sim D$ E-M I

11. $\sim(D \,\&\, \sim D)$ 10 De M

12. $\sim\sim D$ 9, 11 MT

DN then gives us the conclusion we want:

13. D 12 DN

The strategy of indirect proof also comes in handy in proving theorems, though often in such cases it is best to use it in combination with some more ordinary strategy. Thus to establish '$[(C \vee F) \,\&\, (F \supset C)] \supset C$' we might begin with a normal conditional proof, but then nest an indirect one inside it:

1. $(C \vee F) \,\&\, (F \supset C)$ Assumption

  2. $\sim C$ Assumption

  3. $C \vee F$ 1 Simp

  4. F 2, 3 DS

  5. $(F \supset C) \,\&\, (C \vee F)$ 1 Com

  6. $F \supset C$ 5 Simp

  7. C 4, 6 MP

  8. $C \,\&\, \sim C$ 7, 2 Conj

  9. $\sim C \supset (C \,\&\, \sim C)$ 2–8 CP

10. $\sim C \vee \sim\sim C$ E-M I

11. $\sim(C \,\&\, \sim C)$ 10 De M

12. $\sim\sim C$ 9, 11 MT

13. C 12 DN

14. $[(C \vee F) \,\&\, (F \supset C)] \supset C$ 1–13 CP

It has always been annoying to have to display such lines as (5) and (13) above in deductions. So let us agree that from this point on any application of **Com** may be combined with any application of any other rule into a single line in a deduction, provided both **Com** and the other rule are cited as justification for that line; and let's agree to allow ourselves to combine **DN** with any other rule in the same way. Then, for example, the above deduction may be presented in the following more succinct way:[8]

1.	$(C \lor F)$ & $(F \supset C)$	Assumption
2.	$\sim C$	Assumption
3.	$C \lor F$	1 Simp
4.	F	2, 3 DS
5.	$F \supset C$	1 Com, Simp
6.	C	4, 5 MP
7.	C & $\sim C$	6, 2 Conj
8.	$\sim C \supset (C$ & $\sim C)$	2–7 CP
9.	$\sim C \lor \sim\sim C$	E-M I
10.	$\sim(C$ & $\sim C)$	9 De M
11.	C	8, 10 MT, DN
12.	$[(C \lor F)$ & $(F \supset C)] \supset C$	1–11 CP

EXERCISES

I. See if the strategy introduced in this section can be used to construct new deductions for (2), (9), and (10) in group (III) of the Exercises at the end of Section 3.2.

II. Again using the strategy discussed in this section, deduce the conclusions of each of the following arguments from their premises:

1. $A \supset E$
    $B \supset F$
    $(E \lor F) \supset \sim G$
    $(G \lor H) \supset (A \lor B)$
    $\therefore \sim G$

2. $(A \lor L) \lor O$
    $L \equiv O$
    $\therefore O \lor A$

3. $(R \supset S)$ & $(F \supset W)$
    $\sim(S \lor W)$
    $\therefore \sim(F \lor R)$

4. S
    $\therefore (T \supset U) \lor (U \supset T)$

---

[8]When using these short-cut devices, the order in which the rules are cited should reflect the order in which they are applied. Thus, for example at step (5) of the following deduction the citation is displayed as '**Com, Simp**' since **Commutation** is employed first.

5. $A \vee (B \& C)$
   $[D \vee (D \supset E)] \supset$
   $(A \supset C)$
   $\therefore C$

6. $[(A \vee B) \& (C \vee D)] \supset E$
   $\therefore \sim(A \& D) \vee E$

III. Can the strategy of indirect proof be used to construct shorter proofs for (1)–(4) in group (III) of the Exercises at the end of Section 3.3? Explain.

IV. Using whatever methods seem most helpful, show that each of the following arguments is deductively valid:

1. $(C \vee D) \supset (C \supset B)$
   $(\sim A \supset B) \supset \sim(D \vee E)$
   $\sim C \supset D$
   $\therefore (A \vee B) \equiv C$

6. D
   $A \supset B$
   $E \supset C$
   $\sim A \supset (D \supset E)$
   $(B \vee C) \supset F$
   $\therefore F$

2. $A \supset B$
   $B \supset C$
   $\sim A \supset \sim C$
   $(\sim A \& \sim C) \supset F$
   $(A \& C) \supset D$
   $\therefore D \vee F$

7. $(A \supset B) \supset (B \supset D)$
   $O \supset (C \supset S)$
   $(\sim S \supset \sim O) \supset A$
   $\therefore (A \supset B) \supset (\sim C \vee D)$

3. $A \& \sim A$
   $\therefore B$

8. $\sim A \supset [\sim A \supset (B \& B)]$
   $B \supset [A \supset (C \& \sim C)]$
   $\therefore \sim A \equiv B$

4. $(F \supset S) \& (R \supset W)$
   $F \& R$
   $\therefore S \& W$

9. $[(D \& C) \vee (\sim D \& \sim C)] \supset$
      $(S \vee U)$
   $\sim R \supset (D \supset C)$
   $\sim D \supset \sim C$
   $S \supset T$
   $\therefore (R \supset S) \supset (T \vee U)$

5. A
   $\therefore \sim(B \equiv \sim B)$

10. $(\sim\sim A \vee \sim\sim B) \supset (D \supset \sim L)$
    $(C \supset E) \supset [D \& (L \vee \sim A)]$
    $\therefore A \supset C$

11. If Palmer sinks his thirty-foot putt, then if the spectators let out a spontaneous shout then Marr will miss his mere tap-in. If Palmer sinks his thirty-foot putt only if Marr misses his tap-in, one of the two will be on the practice tee tonight. Therefore if the spectators let out a spontaneous shout, or if Marr misses his tap-in, or if Palmer does not sink his thirty-foot putt, then either Palmer or Marr will be on the practice tee tonight.

12. Yossarian is crazy if he flies his missions, and if he is crazy

then he is not obligated to fly his missions. However, if Yossarian asks to be grounded then he is showing concern for his own safety in the face of real, immediate danger, and if he shows such concern then he is *not* crazy and *is* obligated to fly his missions. Moreover, Yossarian won't fly his missions only if he both asks to be grounded and is not obligated to fly them. Therefore Yossarian is not obligated to fly his missions but flies them nonetheless.

*13. If the usual assumptions of intuitive set theory are all true, then the set of all sets that are not members of themselves is a member of the set of all sets that are not members of themselves if and only if the set of all sets that are not members of themselves is not a member of the set of all sets that are not members of themselves. Therefore the usual assumptions of intuitive set theory are not all true.

CHAPTER FOUR

SEMANTIC AND DEDUCTIVE METHODS COMPARED

*4.1 Two Ways of Explicating Validity

When the term 'valid' was introduced in Chapter 1, there were *two* ways of stating, provisionally, what was meant by it:

An argument is valid if and only if its conclusion follows from its premises; that is, if and only if it is impossible for all its premises to be true but its conclusion false.

The tacit assumption was that these were two different ways of saying the same thing, but it should now be apparent that we chose to explicate these suggestions independently.

Thus, in Section 3.1 we concentrated on the notion of one sentence's *following from* certain others. We suggested in effect that so long as our attention is restricted — as it will be throughout this chapter — to arguments whose premises and conclusions are built up from sentence letters by means of the connectives '&', 'v', and '~', the term 'valid' as initially characterized might better be replaced by the more precisely defined term 'deductively valid'.

An argument is **deductively valid** if and only if its conclusion can be deduced from its premises by repeated use of the rules **Conjunction**, **Simplification**, **Addition**, **Disjunctive Syllogism**, **Excluded-Middle Introduction**, **Commutation**, **Association**, **Distribution**, **Double Negation**, and **De Morgan**.

In Sections 2.1 and 2.3, on the other hand, we noticed that truth-tables could be used to determine the possibility of the premises of such an argument being true while its conclusion was false. There the suggestion was, in effect, that the term 'valid' should be replaced with 'truth-table valid'.

An argument is **truth-table valid** if and only if there is no row of its truth-table in which all its premises are assigned the truth-value **T** but its conclusion, the truth-value **F**.

These accounts of validity and of their correlatives, deductive and truth-table invalidity, appear to coincide well with the initial notions of validity and invalidity introduced in Chapter 1. Arguments that (in our technical sense) are deductively valid surely are paradigm cases of those whose conclusions may properly be said to *follow from* their premises; and arguments that are truth-table valid are obviously those for which it is *impossible* for their premises to be true while their conclusions are false. But the deductive methods of Section 3.1 may be directly compared with the semantic methods of Sections 2.1 and 2.3; and though much that is involved in carrying out such a comparison is beyond the scope of an introductory text, the results of that comparison are not.

For one thing, it turns out that the deductive system developed in Section 3.1 is *complete* in the sense that *every truth-table valid argument is deductively valid*[1] — **strong completeness** is the technical term. And the converse is also true — *every deductively valid argument is truth-table valid*. Thus the two methods of our earlier chapters coincide exactly: the truth-table valid arguments are precisely those whose conclusions are deducible, using the rules of Section 3.1, from their premises.

Our system is also complete in another sense — the technical term is **weak completeness**: *every tautology is a theorem*. Again the converse is true — *every theorem is a tautology*. So the sentences obtainable as lines of deductions in which no steps are labeled 'Premise', each being justified instead by **E-M I** or one of our other nine rules, are precisely the sentences that have **T**'s in every row of their truth-tables. As a result, our system is **consistent** in the sense that no contradictions are provable within it, for no contradiction is a tautology. Moreover, from any set of sentences for which there is at least one assignment of truth-values according to which all are assigned **T**'s, no contradiction can be deduced using our rules.

Apart from its intrinsic interest, the exact agreement of these methods has practical consequences as well. If we cannot find a deduction showing an argument to be valid, it need not be that it is invalid, for it may

---

[1]For discussions of this result and those that follow, and for indications of how they may be established, see Section 4.2.

be that we simply lack the necessary inventiveness to find a deduction that will do the job. But the use of truth-tables requires neither ingenuity nor knowledge of strategy. If one understands how truth-tables work, he can mechanically test any argument for validity no matter what its degree of complexity. And in contrast with deductive methods, truth-tables provide a test for *invalidity* as well — one that, though sometimes tedious, can be routinely carried out.

Since the truth-table valid arguments are precisely the deductively valid ones, then, we are free to move back and forth between the two methods. Invalid arguments may be shown to be so by constructing truth-tables. Valid arguments with six or seven sentence letters and requiring truth-tables of 64 or 128 rows may be shown valid instead by the deduction of their conclusions from their premises. Thus with arguments of a substantial complexity, the deductive method has very practical advantages, as well as a certain naturalness that reflects our own intuitive reasoning processes.

EXERCISES

1. Let us call the deductive system developed in Section 3.1 the system **D**. And let **B** be a deductive system just like **D** except for lacking **D**'s rules of *replacement*. **B**'s sole rules, then, are **Conjunction, Simplification, Addition, Disjunctive Syllogism**, and **Excluded-Middle Introduction**. Is **B** consistent? Is **B** strongly complete? Is **B** weakly complete? Explain your reasoning.

*2. Let **C** be a deductive system just like **D** except for lacking the rules **Simplification** and **Disjunctive Syllogism**. Is **C** consistent? Is **C** strongly complete? Is **C** weakly complete?
Hint: Can the left conjunct of any conjunction be deduced from that conjunction using just **C**'s rules? Can the right disjunct of any disjunction be deduced (in **C**) from the latter together with the negation of its left disjunct?

★4.2 Some Metatheorems

In the preceding section we mentioned several results that tie the deductive methods of Section 3.1 to the truth-table methods of Sections 2.1 and 2.3. Though we shall not attempt to establish these results in full detail, we will try to sketch the proofs required. These results are called **metatheorems** because they are claims *about* our deductive system rather than sentences provable within it. We begin with

METATHEOREM I   Every deductively valid argument is truth-table valid.

(converse of S.C.)

To establish this result, we have to show that every conclusion deducible from any premises $\mathscr{P}_1, \ldots, \mathscr{P}_k$ $(k \geq 1)$ must be assigned the truth-value **T** whenever $\mathscr{P}_1, \ldots, \mathscr{P}_k$ are all assigned **T**'s. We establish this by showing that every line, regardless of its line-number, of any deduction in which $\mathscr{P}_1, \ldots, \mathscr{P}_k$ function as premises must be assigned a **T** when they all are.

For the first k lines of such a deduction this is obvious, since those lines will be $\mathscr{P}_1$ through $\mathscr{P}_k$ themselves. And subsequent lines can only be justified by **E-M I** or else be inferred from preceding steps by one of our other nine rules, rules that have already been shown by the truth-table method to lead only from truths to other truths. Thus lines obtained by **E-MI**, being of the form '$p \vee \sim p$', are tautologies (Section 2.1) and so are assigned **T**'s regardless. Those inferred from others by **Conj**, **Simp**, **Add**, or **DS** must have **T**'s when the others do (Section 2.3). And those obtained from preceding lines by **Com**, **Assoc**, **Dist**, **DN**, or **De M** come from them by replacement of an expression with another equivalent to it (Section 2.1); any two such sentences, then, will have the same pattern of **T**'s and **F**'s in their truth-tables so that, in particular, the later one will have a **T** whenever the earlier line does.

Thus deductions begin with premises, and proceed from them at each step only to other sentences that must be true when those already obtained are. If this point is understood, construction of a more rigorous proof is facilitated. We show:

For each positive integer n, any line of a deduction with line-number no greater than n must be assigned a **T** whenever the premises $\mathscr{P}_1, \ldots, \mathscr{P}_k$ of that deduction are assigned **T**'s.[2]

We observe, first, that any line of such a deduction whose line-number is no greater than 1 must be $\mathscr{P}_1$ and so be assigned a **T** when $\mathscr{P}_1, \ldots, \mathscr{P}_k$ all are.

---

[2]Our proof proceeds by mathematical induction (see Appendix C).

Next we let m be any positive integer; we assume that any line of a deduction whose line-number is no greater than m is assigned a **T** when $\mathscr{P}_1,\ldots,\mathscr{P}_k$ are; and we show that all lines of deductions with line-numbers no greater than $m+1$ are assigned **T**'s when the premises, $\mathscr{P}_1$ through $\mathscr{P}_k$, are. Our assumption takes care of lines with numbers between 1 and m, so it is only lines numbered $m+1$ that we need be concerned with, and our discussion above has shown how to handle them. Those that are themselves premises, and those justified by **E-M I**, certainly have **T**'s when the premises do. And those that come from one or two preceding lines by one of our other rules are, as we saw, taken care of by the truth-table considerations of Sections 2.1 and 2.3. Lines inferred from preceding ones by one of these rules must have **T**'s when those earlier lines do, and by our assumption those earlier lines have **T**'s when the premises do.

Now we may conclude that every line of every deduction from the premises $\mathscr{P}_1,\ldots,\mathscr{P}_k$ has a **T** when the latter have **T**'s: the early lines of such deductions automatically have **T**'s in such cases for they *are* just $\mathscr{P}_1$ through $\mathscr{P}_k$; and later lines must have **T**'s when the premises do because our rules, singly or in combination, preserve truth.

It should be obvious that this demonstration of Metatheorem I can be modified to establish

METATHEOREM II   Every theorem is a tautology.   *(converse of w.c)*

After all, the first lines of proofs of theorems can only be justified by **E-M I** and hence must be tautologies; and since later lines must be true when earlier ones are, all lines of all proofs—and so all theorems—must be tautologies as well. But there is an easier way.

Let $\mathscr{T}$ be any theorem. Then any argument with $\mathscr{T}$ as conclusion will be deductively valid. In particular, the argument

$\mathscr{T} \lor \sim\mathscr{T}$
$\therefore \mathscr{T}$

will be deductively valid—we may enter its premise as the first line of our deduction, and use the lines of the proof of $\mathscr{T}$ for the rest. By Metatheorem I, then, the conclusion of this argument must have a **T** in every row of the truth-table in which its premise does. But its premise, being a tautology, has a **T** in *every* row of its truth-table. So $\mathscr{T}$ also has a **T** in every row of its truth-table, that is, $\mathscr{T}$ is a tautology itself!

With our next metatheorem we want to show that we have *enough* rules, that is, that our ten rules are sufficient for deducing *all* tautologies from those which may, by **E-M I**, initiate a deduction:

METATHEOREM III Every tautology is a theorem. *(weak completeness)*

It is helpful to introduce some temporary terminology. By a **literal** we shall mean a sentence letter or the negation of a sentence letter. Thus 'A', 'B', 'C', and so on are literals, as are '$\sim A$', '$\sim B$', '$\sim C$', and so on. But '$\sim\sim A$' is not (too many curls), and neither are '$A \mathrel{\&} B$' and '$B \vee C$'. We shall be especially interested in literals, disjunctions of literals, and conjunctions of such disjunctions:

DEFINITION A sentence is in **conjunctive normal form (CNF)** if and only if it is either (i) a literal, or (ii) a disjunction of literals, or (iii) a conjunction each of whose conjuncts is either literal or a disjunction of literals.

Thus 'A' and '$\sim B$' are in CNF by clause (i); so are '$A \vee \sim B$' and '$\sim A \vee [(B \vee \sim C) \vee D]$', by clause (ii); and so is '$(\sim A \vee B) \mathrel{\&} [(A \vee \sim D) \mathrel{\&} (B \vee C)]$', by clause (iii). But '$\sim(A \vee B)$', '$\sim\sim A \vee (B \vee \sim C)$', '$(\sim A \mathrel{\&} B) \vee \sim\sim C$', and '$(A \mathrel{\&} B) \vee C$' are not.

To establish Metatheorem III one needs five sub-results or **lemmas**. In our discussion of each we shall omit certain details that a full-fledged proof would call for.[3]

Lemma 1 For each sentence $\mathcal{T}$ there exists a corresponding sentence $\mathcal{T}*$ in conjunctive normal form that is interdeducible with $\mathcal{T}$, that is, that can be deduced from $\mathcal{T}$ using our ten rules and from which $\mathcal{T}$ can be deduced.

To build such a sentence $\mathcal{T}*$ from any given sentence $\mathcal{T}$ we can follow a set procedure: at Stage One drive all curls inside as far as possible by repeated uses of **De M** until no curls occur outside parentheses but only inside, in front of sentence letters; at Stage Two, use **DN** as often as necessary to pare down to no more than *one* curl in front of any sentence letter; then, at Stage Three, use **Com** and **Dist** repeatedly to change parts of the sentence of the form '$p \vee (q \mathrel{\&} r)$' or '$(q \mathrel{\&} r) \vee p$' to the form '$(p \vee q) \mathrel{\&} (p \vee r)$', until we wind up with a conjunction of disjunctions of literals.

---

[3]For amplification, see D. Hilbert and W. Ackermann, *Principles of Mathematical Logic*, New York: Chelsea, 1950, pp. 11–16, whose exposition we follow fairly closely.

Suppose, for example, that $\mathcal{T}$ is the sentence

~{[*B* v ~(*C* & ~*D*)] & *A*}.

We first use **De M** again and again to drive all the curls that come up as we go along as far inside as possible at the first stage.

STAGE ONE: ~[*B* v ~(*C* & ~*D*)] v ~*A*
              [~*B* & ~~(*C* & ~*D*)] v ~*A*
              [~*B* & ~(~*C* v ~~*D*)] v ~*A*
              [~*B* & (~~*C* & ~~~*D*)] v ~*A*

Then we use **DN** again and again until no more than one curl occurs in front of any letter.

STAGE TWO: [~*B* & (*C* & ~~~~*D*)] v ~*A*
              [~*B* & (*C* & ~*D*)] v ~*A*

Finally, we commute parts of the form '(*q* & *r*) v *p*' as they arise, and then do distributions as often as we have to in order to wind up with a conjunction of disjunctions (of literals). This is the final stage.

STAGE THREE: ~*A* v [~*B* & (*C* & ~*D*)]
               (~*A* v ~*B*) & [~*A* v (*C* & ~*D*)]
               (~*A* v ~*B*) & [(~*A* v *C*) & (~*A* v ~*D*)]

The sentence in CNF that we finish with is the sentence $\mathcal{T}*$ that we want. Since **De M**, **DN**, **Com**, and **Dist** are all "reversible" rules, we can be sure not only that $\mathcal{T}*$ is deducible from $\mathcal{T}$, but also that $\mathcal{T}$ is deducible from $\mathcal{T}*$ — to construct the deduction, one need only *reverse* the above steps.

Lemma 2   A sentence is a tautology if and only if each sentence interdeducible with it is a tautology.

Let $\mathcal{T}$ and $\mathcal{T}*$ be any two sentences that are interdeducible, that is, such that $\mathcal{T} \vdash \mathcal{T}*$ and $\mathcal{T}* \vdash \mathcal{T}$. Since every deductively valid argument is, according to Metatheorem I, truth-table valid, then $\mathcal{T} \models \mathcal{T}*$ and $\mathcal{T}* \models \mathcal{T}$. So $\mathcal{T}*$ must be assigned the value **T** whenever $\mathcal{T}$ is, and conversely. If $\mathcal{T}$ is a tautology, then, $\mathcal{T}*$ must also be; and if $\mathcal{T}*$ is a tautology so is $\mathcal{T}$.

Lemma 3   If a sentence in CNF is a tautology, then each of its conjuncts (if any) must be a tautology also.

To see that this is so, recall that conjunctions only have **T**'s in those rows of their truth-tables in which *all* their conjuncts have **T**'s. If one of the conjuncts of a tautologous conjunction were not itself a tautology, then, it would have the truth-value **F** in at least one row of its truth-table. But then the whole conjunction would have an **F** in that row, also, and not be a tautology at all.

Lemma 4   If each conjunct of a sentence in CNF is a tautology, then each conjunct of that sentence must be a disjunction of literals that contains for some sentence letter both that letter and its negation.

To establish this lemma, notice that if some conjunct of such a sentence did *not* contain, for at least one sentence letter, *both* that letter *and* its negation, then that conjunct would have an **F** in the row of its truth-table in which each unnegated letter occurring in it had the truth-value **F** and each negated one the value **T**, and so would not be a tautology. (No conjunct of such a sentence could be a literal, then, for no literal is a tautology.)

Lemma 5   All disjunctions of literals that contain for some sentence letter both it and its negation are theorems.

In fact it is easy to see that any such disjunction can be proved by starting with **E-M I**, using **Add** to get in the other disjuncts required, and then using **Com** and **Assoc** to get the order and grouping right.

Once lemmas 1–5 have been established, we can return to Meta-theorem III and show that every tautology is a theorem. For let $\mathscr{T}$ be any tautology. By lemma 1 we can find a sentence $\mathscr{T}*$ in CNF that is inter-deducible with $\mathscr{T}$. Since $\mathscr{T}$ is a tautology, lemma 2 assures us that $\mathscr{T}*$ is a tautology also. But then lemma 3 tells us that each of $\mathscr{T}*$'s conjuncts must be a tautology as well. It then follows from lemma 4 that each conjunct of $\mathscr{T}*$ is a disjunction of literals which contains for some sentence letter both that letter and its negation. By lemma 5 we can conclude that each of $\mathscr{T}*$'s conjuncts is a theorem. But then $\mathscr{T}*$ must also be a theorem—we can prove each of its conjuncts (since they're all theorems) and then use **Conj** to stick them together to get $\mathscr{T}*$. Now according to lemma 1, $\mathscr{T}$ is deducible from $\mathscr{T}*$, so if we continue the proof of $\mathscr{T}*$ by proceeding to deduce $\mathscr{T}$ from *it*, we finish with a proof of $\mathscr{T}$. $\mathscr{T}$ must, then, be a theorem.

To see how the whole process works, let's go through one example in detail. Suppose $\mathcal{T}$ is the tautology '$[\sim(\sim B\ \&\ B)\ \&\ (D\ \mathrm{v}\ \sim\sim\sim C)]$ v C'. We first find a sentence $\mathcal{T}*$ in CNF that is interdeducible with $\mathcal{T}$ by following the procedure described for lemma 1:

$\mathcal{T}$	$[\sim(\sim B\ \&\ B)\ \&\ (D\ \mathrm{v}\ \sim\sim\sim C)]$ v C	
	$[(\sim\sim B\ \mathrm{v} \sim B)\ \&\ (D\ \mathrm{v}\ \sim\sim\sim C)]$ v C	(De M)
	$[(B\ \mathrm{v} \sim B)\ \&\ (D\ \mathrm{v}\ \sim\sim\sim C)]$ v C	(DN)
	$[(B\ \mathrm{v} \sim B)\ \&\ (D\ \mathrm{v} \sim C)]$ v C	(DN)
	C v $[(B\ \mathrm{v} \sim B)\ \&\ (D\ \mathrm{v} \sim C)]$	(Com)
$\mathcal{T}*$	$[C\ \mathrm{v}\ (B\ \mathrm{v} \sim B)]\ \&\ [C\ \mathrm{v}\ (D\ \mathrm{v} \sim C)]$	(Dist)

As lemmas 2–4 promised, $\mathcal{T}*$ is also a tautology, and since it is in conjunctive normal form, each of its conjuncts is a tautology as well, and so each is a disjunction of literals containing for some sentence letter both that letter and its negation. By lemma 5 we can consequently prove each of its conjuncts, and then conjoin them to obtain $\mathcal{T}*$:

1.	B v $\sim B$	E-M I
2.	$(B$ v $\sim B)$ v C	1 Add
3.	C v $(B$ v $\sim B)$	2 Com
4.	C v $\sim C$	E-M I
5.	$(C$ v $\sim C)$ v D	4 Add
6.	C v $(\sim C$ v $D)$	5 Assoc
7.	C v $(D$ v $\sim C)$	6 Com
8.	$[C$ v $(B$ v $\sim B)]\ \&\ [C$ v $(D$ v $\sim C)]$	3, 7 Conj

To obtain our proof of $\mathcal{T}$, then, we have only to extend this proof of $\mathcal{T}*$ by reversing the steps we went through before in locating it:

9.	C v $[(B$ v $\sim B)\ \&\ (D$ v $\sim C)]$	8 Dist
10.	$[(B$ v $\sim B)\ \&\ (D$ v $\sim C)]$ v C	9 Com
11.	$[(B$ v $\sim B)\ \&\ (D$ v $\sim\sim\sim C)]$ v C	10 DN
12.	$[(\sim\sim B$ v $\sim B)\ \&\ (D$ v $\sim\sim\sim C)]$ v C	11 DN
13.	$[\sim(\sim B\ \&\ B)\ \&\ (D$ v $\sim\sim\sim C)]$ v C	12 De M

With this work behind us, the proof of our last metatheorem is relatively simple:

METATHEOREM IV   Every truth-table valid argument is deductively valid. *(strong complet.)*

Suppose that the argument

$$\mathscr{P}_1$$
$$\vdots$$
$$\mathscr{P}_k$$
$$\therefore \mathscr{C}$$

is truth-table valid. Then there can be no row of its truth-table in which $\mathscr{P}_1$ through $\mathscr{P}_k$ are all assigned **T**'s but $\mathscr{C}$ an **F**. Forgoing internal parentheses in extended conjunctions, we consider

$$\sim(\mathscr{P}_1 \,\&\, \ldots \,\&\, \mathscr{P}_k) \vee \mathscr{C},$$

which clearly must be a tautology — any row in which it had the value **F** would have to be a row in which $\mathscr{P}_1$ through $\mathscr{P}_k$ were assigned **T**'s but $\mathscr{C}$ an **F**, and there are no such rows. Since all tautologies are theorems, then, it is a theorem. And now we can see that the argument in question must be deductively valid, for the deduction of $\mathscr{C}$ from $\mathscr{P}_1, \ldots, \mathscr{P}_k$ could proceed as follows. First, use **Conj** repeatedly on the premises until

$$\mathscr{P}_1 \,\&\, \ldots \,\&\, \mathscr{P}_k$$

has been deduced. Then use **DN** to obtain

$$\sim\sim(\mathscr{P}_1 \,\&\, \ldots \,\&\, \mathscr{P}_k).$$

Then insert the proof guaranteed us by Metatheorem III of the theorem

$$\sim(\mathscr{P}_1 \,\&\, \ldots \,\&\, \mathscr{P}_k) \vee \mathscr{C}.$$

One last step, using **DS**, then delivers $\mathscr{C}$!

EXERCISES

1. Let us call the deductive system developed in Section 3.1 the system **D**. Students sometimes wonder why we bother to study both **D** and the truth-table methods of Chapter 2 when they turn out to be, in effect, merely two different ways of doing one and the same thing. One reason for studying both methods is that often one of them affords the simplest way to answer a question about the other. Consider, for example, the question "If $\mathscr{P}$ and $\mathscr{Q}$ are sentences built up from letters and the three connectives '&', 'v', and '~', and $\mathscr{Q}$ is deducible (using **D**'s ten rules of inference and replacement) from $\mathscr{P}$, must it be the case that the disjunction

   $$\sim\!\mathscr{P} \text{ v } \mathscr{Q}$$

   is a theorem of **D**?" For this exercise, show that the answer is "Yes." You will probably find it helpful to establish this result concerning **D** by making considerable use of what you know about truth-table methods, along with the connections between deductive and semantic methods discussed above.

*2. Let **D′** be the deductive system developed in Sections 3.1 and 3.2 for treating sentences built up from sentence letters and the *five* connectives '&', 'v', '~', '⊃', and '≡'. Extend the treatment of **D** presented in this chapter to **D′** by redefining such terms as 'deductively valid' and 'truth-table valid' so as to take into account the additional connectives dealt with in **D′**. Then state analogues for **D′** of Metatheorems I–IV, and explain how each might be established.

*3. When the rule of conditional proof was introduced in Section 3.3 we said that it was not really needed, that we could not deduce any conclusions from any premises *with* that rule that we could not have deduced *without* it. Let **D″** be the deductive system developed in Sections 3.1–3.4, and use the work you've done in the two preceding exercises to justify that earlier claim by showing that any argument whose conclusion can be deduced from its premises using the rules of **D″** is such that its conclusion can also be deduced from its premises using just the rules available in **D′**.

CHAPTER FIVE

QUANTIFICATIONAL LOGIC
Symbolism and Semantics

5.1 Quantifiers and Related Apparatus

While the techniques developed in Chapters 2 and 3 have a wide application, there are nonetheless many arguments that fall outside their scope. Even a simple example like

Every philosopher is wise
Aristotle is a philosopher
Therefore Aristotle is wise

cannot be handled. For since the premises and conclusion are all simple sentences containing no others as parts, the argument is of the patently invalid form

p
q
$\therefore r$

Yet of course it is valid, and to show that it is, we need to focus on the internal composition of the premises and conclusion—in particular, on the relative positions of the terms 'philosopher', 'wise', and 'Aristotle'. Only in this way shall we be able to distinguish valid arguments like the above from invalid ones like

Every philosopher is wise
Aristotle is wise
Therefore Aristotle is a philosopher.

To develop techniques for establishing validity and invalidity in such cases, we must first examine the internal structure of uncompounded sentences such as 'Every philosopher is wise', 'Aristotle is a philosopher', and the like. This is the main subject of the present chapter; techniques for validity will be taken up in the next.

Terms like 'Aristotle' are typically used in a given context to refer to a single individual and hence are called **singular terms**. All proper names (such as 'Fido', '3', 'Indiana', 'Nixon') fall into this category, and while their reference might vary from one context to another—as with 'John', for example—nonetheless, once the context has been fixed, such terms serve to single out one specific thing or individual for discussion. In addition to proper names, there is another class of singular terms known as *definite descriptions*, examples of which include 'the author of *De Anima*', 'the thirty-seventh President of the United States', and 'John's mother'. Such terms do not name an individual but rather purport to uniquely *describe* one. Aristotle was not *named* 'the author of *De Anima*' but 'Aristotle'; had he not actually written *De Anima* that description would not apply to him. Sentences containing definite descriptions require a rather complex analysis, however, so we shall postpone discussion of them until Section 5.4.

Unlike 'Aristotle', such terms as 'philosopher' and 'wise' do not serve to refer, either by name or description, to one particular individual. Instead, they are essentially classificatory in function; they typically represent *properties* that a number of individuals might share—in the present case the properties of being a philosopher and of being wise. We shall call such expressions **general terms** since they may characterize or describe a general class of individuals (those having the property in question) rather than referring to just one. Thus 'philosopher' is a general term in our sense—while it does not refer to any specific individual, it describes such individuals as Aristotle and Plato.[1]

General terms may occur in noun, verb, and adjectival form, for example 'smoker' in 'Al is a smoker', 'smokes' in 'Al smokes', and so on. And though they are classificatory and descriptive, it must not be thought that general terms must *in fact* describe any actual objects. 'Unicorn' is a general term, for example, though the property of unicornhood is possessed by no individual.

A sentence typically used to ascribe a property to a specifically designated individual will be called a **singular** sentence. The grammatical subject of such a sentence is a singular term and the predicate contains a general term. The second premises and the conclusions of the arguments

---

[1] Any definite description will contain a general term, for example 'author' in 'the author of *De Anima*'. But because it contains the definite article 'the' in the singular, the definite description *itself* serves not to characterize a class of individuals, but to refer to just one. It is, therefore, a singular term.

cited earlier, for example, are singular sentences. It will be convenient also to regard truth-functional compounds of singular sentences as themselves singular. So 'Bob and Mary are students' and 'Bob is either married or divorced' count as singular sentences as well. And sentences like 'Chicago is larger than Detroit', in which a *relation* is said to hold between two or more specified individuals (instead of a property ascribed to just one), will also be considered singular. The latter will be taken up in Section 5.2; here we concentrate on those in which properties are ascribed to individuals.

The first premise of our sample argument, 'Every philosopher is wise', does *not* serve to ascribe a property to one or more specified individuals, and thus is not a singular sentence. Rather, it says in effect that any individual that has the property of being a philosopher also has the property of being wise, but no specific individual — whether philosopher or no — is referred to. Such a sentence will be characterized as **general**, for its grammatical subject and predicate each contain general terms with the subject usually containing in addition a *quantity* expression such as 'every', 'all', or 'some'. General sentences may also contain singular terms, but, as the first premise illustrates, they need not do so. Unlike the singular variety, a general sentence relates together classes of individuals — in the present case, philosophers and wise persons.

We may sharpen this somewhat rough distinction between singular and general sentences as follows. With a singular sentence like 'Aristotle is wise', there is no distinction in meaning if we negate it "externally" ('It is not the case that Aristotle is wise') or "internally", by the insertion of 'not' into the predicate ('Aristotle is not wise'). The resulting sentences are synonymous, and both might be translated, using our old notation, as '$\sim A$'. But a typical general sentence like 'Every philosopher is wise' yields quite different results depending on how 'not' is attached. Externally, we arrive at its genuine negation:

It is not the case that every philosopher is wise

(or, more idiomatically, 'Not every philosopher is wise'). And this is equivalent to

Some philosopher is not wise.

But placing 'not' into the predicate gives

Every philosopher is not wise

which, understood literally, is equivalent to

No philosopher is wise.

Clearly the latter is *not* the negation of 'Every philosopher is wise', for both of these may well be false. Only when 'not' is applied externally to a general sentence do we derive a sentence that may properly be described as its negation.[2]

Now if we consider the singular sentences:

(1) Aristotle is a philosopher

(2) Brutus is a philosopher

(3) Cassius is a philosopher

we may abstract from them the schematic predicate expression '...is a philosopher' common to all three. And this expression in its entirety, rather than the component general term, may be taken as representing the property of being a philosopher. Our practice will thus be to treat general terms as integral parts of predicative contexts in which they occur.

We shall use the capital letters '*A*', '*B*', '*C*',... to abbreviate such predicate expressions. These letters thus represent properties of individuals, in effect, and we shall call them **predicate letters**.[3] To serve as proper names within our notation, we shall use the small letters '*a*' through '*t*'. They will serve to designate specific individuals just as 'Aristotle' and 'Brutus' ordinarily do, and because their designations will be fixed in any given context in which we use them, they will be called **individual constants**.

With these notational devices we have the means for expressing singular sentences in compact style. Thus, where '*a*', '*b*', and '*c*' designate Aristotle, Brutus, and Cassius respectively, (1), (2), and (3) become:

(1') *Pa*

(2') *Pb*

(3') *Pc*

---

[2]Note that some sentences that are singular in form are often used like general sentences. For example, 'The whale is a mammal' may be used not to refer to any particular whale, but rather to the entire species. As such, it has the force of 'All whales are mammals'.

[3]It is perhaps worth emphasizing that predicate letters are not mere stand-ins for general terms, but rather abbreviate entire predicative contexts of which general terms are usually but a part. Often these contexts are rather complex, and they may even include singular terms (for example, '... conspired with Cassius').

Small letters from the end of the alphabet, '*w*', '*x*', '*y*', '*z*', '*x''*', '*y''*', etc. will be employed as **individual variables**.[4] They will not be used to designate specific individuals, but rather will range indiscriminately over all of them. Consequently expressions like

(4) *x* is a philosopher

and

(4') *Px*

are neither true nor false, since no particular individual is designated by '*x*', and thus neither is a sentence. (4') is related to (1') much as '$x^2 = 9$' is to '$3^2 = 9$', and both (4') and its English counterpart (4) will be called **open sentences**. Such expressions stand to genuine sentences much as counterfeit money stands to the real thing. An open sentence is not itself either true or false, as a genuine sentence would be. But an open sentence may be transformed *into* a real sentence by replacing the variables in it with individual constants. Hence a singular sentence like '*Pa*' will be termed a **substitution instance** of the open sentence '*Px*', since it is derived from the latter through such replacement. An open sentence, then, must contain at least one individual variable that is subject to replacement by a constant.

Now we can turn to the task of expressing general sentences. The use of such sentences in any given context presupposes at least implicitly a definite *universe of discourse* — the range of individuals under discussion in that context. The universe may range anywhere from the actual universe to something much smaller. When an instructor looks out over his class and says "Everyone is here," for example, he clearly does not mean every person *whatever* but rather every person enrolled in his course. Were we to object that his remark is false because, say, the British Prime Minister is not there, the objection would hardly be entertained seriously. The universe of discourse is tacitly restricted to enrolled persons, and hence the sentence would be counted as false only if at least one enrolled person were not in attendance.

It will be helpful in examining general sentences to restrict ourselves initially to a small and artificial universe of discourse. We shall therefore consider a universe with three individuals *a*, *b*, and *c*, together with five properties which we shall represent by the letters '*F*', '*G*', '*H*', '*J*', and '*K*'. (We need not worry here about *which* specific individuals and properties they might be.) Now whether a sentence is true or false in this universe will depend on which individuals exhibit which properties. We

---

[4]The unused letters '*u*', '*v*', '*u''*', '*v''*', etc. are being reserved for a special use in Chapter 6.

shall suppose that they are distributed as follows (a '+' means that the individual has the property, a '−' that it lacks it):

	F	G	H	J	K
a	+	−	+	−	−
b	−	+	+	−	+
c	+	+	+	−	−

A glance at this chart will show that the following singular sentences (among others) are *true*:

Fa
~Ga
Fb v Hb
Fa & ~Jc

Moreover, each of the following general sentences (stated for the moment in English phraseology) are true:

(5) Something is F

(6) Everything is H

(7) Nothing is J

(8) Something is not G

(9) All F are H

(10) No F are J

(11) Some F are G

(12) Some F are not G

It may seem tempting at first to express (5) as '*Fs*' where '*s*' means 'something'. But a little reflection will show this to be absurd. To use '*s*' as an individual constant presupposes that it names or designates a specific individual, but there is clearly no individual named 'something'. An exhaustive enumeration of the individuals of our universe would include just *a*, *b*, and *c*. There is no fourth dubbed 'something', much less a fifth named 'nothing' to which the property *J* (whatever it may be) is ascribed in (7). Moreover, (5) admits of distinct "negations" that the expression '*Fs*'—admitting only the single negation '~*Fs*'—cannot handle. For 'Something is not *F*' is *true* in our universe (since *b* does not have *F*) but 'It is not the case that something is *F*' is *false* (*a* and *c* both have *F*).

It will be better to approach (5) from a different angle. Clearly (5) will be false just in case *none* of the individuals in our universe of discourse have *F*, and therefore will be *true* just in case *there is at least one* that does. Thus 'some' here means 'there is at least one'. And because our universe has so few individuals in it, we can reformulate the point. Since 'Something is *F*' is true just in case there is at least one thing that is *F*, that is, true if and only if either *a*, *b*, or *c* is *F*, then it is true just in case in least one substitution instance of the open sentence '*Fx*' is true.[5] And since '*Fa*' and '*Fc*' are both true, (5) is also.

(5), then, is most fully expressed as

There is at least one individual such that it is *F*

and, since the variable '*x*' ranges over individuals, this in turn may be reexpressed as

There is at least one *x* such that *x* is *F*.

Now '*x* is *F*' is rendered in symbols '*Fx*', and we saw that (5) is true just in case at least one individual in the universe of discourse has the property *F*. We shall express a sentence having such a truth-condition by attaching the expression '$(\exists x)$' to '*Fx*', and hence express (5) as

(5') $(\exists x)(Fx)$.

The sign '$(\exists x)$' may be read 'there is at least one *x* such that...' (or more briefly 'for some x,...') and is called an **existential quantifier**.

It should be noted that the second occurrence of '*x*' in (5') is no longer subject to replacement by a constant, for attaching '$(\exists x)$' to '*Fx*' has yielded not another open sentence but a genuine sentence, one that is the translation of 'Something is *F*' and so is either true or false. The role of the second '*x*' in (5') parallels that of the pronoun 'it' in the first of the above reformulations of (5), that is, it carries cross-reference to the first '*x*' just as 'it' carries cross-reference to 'individual'. Clearly '*y*' or '*z*' would have done as well provided that the quantifier had been altered appropriately — '$(\exists y)(Fy)$' and '$(\exists z)(Fz)$' would also do as translations of (5).

Turning now to (6), we notice that it is true because every individual

---

[5]This alternate interpretation of (5) is known as the *substitution* interpretation. We use it only as a pedagogical device, and only here, where the universe of discourse under discussion is finite. When more realistic universes are considered, such as the set of real numbers, we shall drop this mode of speech.

has the property H; and this may be put by saying that (6) is true in our universe of discourse just in case each substitution instance of the open sentence 'Hx' is true. Since all of 'Ha', 'Hb', and 'Hc' are true, then, (6) is too. So (6) may be more fully expressed as

For any individual whatever, it is H

or as

For any x, x is H.

We shall express a sentence that is true just in case each and every individual in the universe of discourse has the property H by attaching the expression '(x)' to the open sentence 'Hx'. Thus (6) is expressed in symbols by

(6') $(x)(Hx)$.

The sign '(x)' may alternately be read as 'given any x,...' or 'for all x,...'. We shall call it a **universal quantifier**.

By the same token, (7) is true because none of the individuals in our universe has J. That is, (7) is true in our universe just in case all substitution instances of 'Jx' are false, and this means, of course, when all substitution instances of '$\sim Jx$' are *true*. So we may express (7) by attaching the universal quantifier to the open sentence '$\sim Jx$':

(7') $(x)(\sim Jx)$.

Since (7) is true just in case *no* substitution instance of 'Jx' is true, it follows that '$(\exists x)(Jx)$' is false just in case (7) is true. Hence (7) may also be expressed as the *negation* of '$(\exists x)(Jx)$',

(7'') $\sim(\exists x)(Jx)$,

whose English counterpart would be 'There is no individual that has J'. More generally, where

...x...

represents any open sentence containing 'x', sentences of the forms:

$(x)(\sim(\ldots x\ldots))$ and $\sim(\exists x)(\ldots x\ldots)$

are equivalent.[6]

Turning now to (8), its truth is apparent from the fact that at least one individual lacks the property G. Hence at least one substitution instance of '$\sim Gx$' is true, and accordingly the existential quantifier may be employed:

(8') $(\exists x)(\sim Gx)$.

But this is no different from saying that (8) is true just when *not all* things are G. Therefore (8) may alternately be expressed using the universal quantifier as

(8'') $\sim(x)(Gx)$

which goes into English as 'Not everything is G'. So sentences of the forms

$(\exists x)(\sim(\ldots x\ldots))$ and $\sim(x)(\ldots x\ldots)$

are equivalent.

The sentences (9) through (12) are respectively of the forms:

A: all ... are ---
E: no ... are ---
I: some ... are ---
O: some ... are not ---.

The labels '**A**', '**E**', '**I**', and '**O**' were originally applied to these forms in the traditional logic dating from Aristotle, and were regarded as basic to

---

[6]Parentheses are shown in the left-hand schema as a reminder that they are necessary in the case of a typical substitution for $\ldots x\ldots$ such as '$Fx \,\&\, Gx$' — without them, we would have '$(x)(\sim Fx \,\&\, Gx)$' instead of the desired '$(x)(\sim(Fx \,\&\, Gx))$'. Of course when the substitution for $\ldots x\ldots$ is not compound (e.g. 'Fx') or has a curl as main connective (e.g. '$\sim Gx$' or '$\sim(Fx \,\&\, Gx)$') the inner parentheses are superfluous and should be mentally erased from the above schema. The same considerations apply to similar schemata displayed subsequently in the text.

nearly all inferences involving general sentences. With the development of modern symbolic logic, they have come to be recognized as less fundamental than earlier logicians had supposed, but they still provide us with an excellent point of departure for the study of general sentences.

(9)—'All *F* are *H*'—is an **A**-sentence and such sentences can obviously take variant forms:

every ... is ---
any ... is ---
each ... is ---
...'s are all ---.

When we consider how an **A**-sentence like (9) should be expressed, it might seem natural at first to try

(*x*)(*Fx* & *Hx*).

But a moment's reflection will show that this will not do, for it is *false* in our universe, whereas (9) is obviously true. Since '*Fb*' is false, the open sentence '*Fx* & *Hx*' has the false substitution instance '*Fb* & *Hb*', and this suffices to make '(*x*)(*Fx* & *Hx*)' false as well. It in fact asserts something far stronger than (9), for it says that everything is *both F* and *H* (compare 'All men are mortal' with 'Everything (whatever) is both a man and mortal'). On the other hand, it is true of each individual in our universe that *if* it is *F*, then it is also *H*; accordingly, all substitution instances of '*Fx* ⊃ *Hx*' are true. For '*Fa* ⊃ *Ha*' and '*Fc* ⊃ *Hc*' have both a true antecedent and a true consequent, and '*Fb* ⊃ *Hb*' has a false antecedent. '*Fx* ⊃ *Hx*' can have a false substitution instance only when the antecedent is true and the consequent false, and this clearly matches up with the intended meaning of (9); it will be false, that is, if at least one individual has *F* but does *not* have *H*. Thus (9) is more fully expressed in the conditional form

For any *x*, if *x* is *F* then *x* is *H*

and hence may be translated as

(9′) (*x*)(*Fx* ⊃ *Hx*).

While (9′) is true in our universe, '(*x*)(*Fx* ⊃ *Gx*)' is not, for '*Fx* ⊃ *Gx*' admits of the false substitution instance '*Fa* ⊃ *Ga*'.

Now the **E**-sentence (10) — 'No F are J' — is true because there is no individual that has F and *also* has J. Accordingly, the open sentence '$Fx \supset \sim Jx$' has all true substitution instances. Both antecedent and consequent are true in '$Fa \supset \sim Ja$' and '$Fc \supset \sim Jc$', and '$Fb \supset \sim Jb$' has a false antecedent. So (10) says in effect that each and every F is *not* J and may be expressed as

(10′) $(x)(Fx \supset \sim Jx)$.

But 'No F are G' is false in our universe, since c has both F and G.

In view of the foregoing, it is tempting to translate the **I**-sentence (11) — 'Some F are G' — as:

$(\exists x)(Fx \supset Gx)$.

However, it is easily shown that this is incorrect. Consider the **I**-sentence, 'Some F are K'; this is clearly false in our universe since no individual that has F also has K. Yet '$(\exists x)(Fx \supset Kx)$' is *true* (the open sentence '$Fx \supset Kx$' has at least one true substitution instance, '$Fb \supset Kb$') so it cannot be the correct rendering of 'Some F are K'. By the same token '$(\exists x)(Fx \supset Gx)$' fails to correctly translate (11). It is too weak a claim, in that it will be true whenever there is at least one individual that is *not* F, and (11) obviously requires for its truth that there be at least one individual that *is* F and moreover is G. This means that (11) will be true just in case '$Fx \& Gx$' has at least one true substitution instance — which it does, since '$Fc \& Gc$' is true. Thus (11) becomes 'There is at least one x such that x is F and x is G', which goes into symbols as

(11′) $(\exists x)(Fx \& Gx)$.

So whereas 'All F are G' is translated with '$\supset$' instead of '$\&$', the situation is just the reverse with 'Some F are G' — we should not be misled by the occurrence of 'are' in both. Note also that **E**- and **I**-sentences are contradictories — the negation of the **E**-sentence is equivalent to the corresponding **I**-sentence and vice versa. For '$(x)(Fx \supset \sim Gx)$' will be true just in case *no* individual has both F and G, whereas (11′) will be true just in case at least one *does* have both F and G. Thus, the following pairs are equivalent forms:

$(\exists x)(\ldots x \ldots \& ---x---)$ and $\sim(x)(\ldots x \ldots \supset \sim(---x---))$
$(x)(\ldots x \ldots \supset \sim(---x---))$ and $\sim(\exists x)(\ldots x \ldots \& ---x---)$

The **O**-sentence (12) — 'Some *F* are not *G*' — may be handled on the model provided by (11). Thus it is true just in case at least one individual has *F* but lacks *G*. Accordingly, the open sentence '*Fx* & ~*Gx*' must have at least one true substitution instance — which in fact it does: '*Fa* & ~*Ga*'. We thus have

(12′) (∃*x*)(*Fx* & ~ *Gx*).

O-sentences and **A**-sentences are also contradictories, for (12′) will be true just in case at least one individual has *F* but not *G*, whereas '(*x*)(*Fx* ⊃ *Gx*)' will be true if and only if *no* individual is such that it has *F* but lacks *G*. Thus we obtain the equivalent pairs:

(∃*x*)(...*x*... & ~(---*x*---))   and   ~(*x*)(...*x*... ⊃ ---*x*---)
(*x*)(...*x*... ⊃ ---*x*---)   and   ~(∃*x*)(...*x*... & ~(---*x*---))

Moving beyond the artificial universe of discourse now, let us consider sentences of somewhat greater complexity. The sentence

Some Republicans are either liberal or moderate

goes readily into symbols as

(∃*x*)[*Rx* & (*Lx* v *Mx*)].

Parentheses are important, for '(∃*x*)[(*Rx* & *Lx*) v *Mx*]' makes the quite different claim that something either is a Republican liberal or else is moderate. Many sentences containing no quantity expression like 'all' or 'some' nonetheless require quantifiers in their translation, as can be seen with sentences beginning with a plural noun. However, a consideration of the normal context of utterance is usually required to discern whether the quantifier should be universal or existential.

  Compare, for example, 'Horses are mammals' and 'Horses are coming down the road'. The first would customarily be used to make a claim about the entire species and hence should be expressed as the **A**-sentence '(*x*)(*Hx* ⊃ *Mx*)'. But the second generally would not be taken as an assertion about *all* horses, whatever the presumed universe of discourse. The meaning here is that of 'some' and so 'Horses are coming down the road' should be expressed as '(∃*x*)(*Hx* & *Cx*)'. Similar considerations hold for sentences beginning with the indefinite article. We have the same contrast as before between 'A horse is a mammal' and 'A horse is coming down the road', with the same respective translations required.

Sentences containing 'only' can be especially troublesome. Those beginning with this word usually require for translation inverting the position of the component general terms. Thus the true sentence

Only men are eligible for the draft

should be rendered as

$(x)(Ex \supset Mx)$,

not as '$(x)(Mx \supset Ex)$', which would make the much different (and false) claim that *every* man is eligible for the draft.

The sentence 'Cats and dogs are vertebrates' may be translated as

$(x)[(Cx \lor Dx) \supset Vx]$

or equivalently as

$(x)(Cx \supset Vx) \& (x)(Dx \supset Vx)$.

But it must *not* be translated '$(x)[(Cx \& Dx) \supset Vx]$', for this says that anything that is *both* a cat and a dog is a vertebrate — hardly the intended meaning. On the other hand, the sentence

Any cat that howls at night is a nuisance

may be translated (using 'Hx' for 'x howls at night') with an '$\&$' in the antecedent:

$(x)[(Cx \& Hx) \supset Nx]$.

Some sentences raise problems concerning their main connective. Consider for example

Some men achieve greatness if they work hard.

This says that there are men who, if they work hard, achieve greatness. Accordingly, we have

$(\exists x)[Mx \ \& \ (Wx \supset Gx)]$

where 'Wx' is 'x works hard' and 'Gx' is 'x achieves greatness'. We should not translate it as '$(\exists x)[(Mx \ \& \ Wx) \supset Gx]$', for this, unlike the English sentence, would be true merely if no individual is a hard-working man (because then any substitution instance of the component open sentence would be true). The English sentence does not say that there are individuals that, if they are hard working men, achieve greatness; it says that there are individuals that are men *and* who are such that they achieve greatness if they work hard. So '&' is the main connective, not '$\supset$'. The sentence

All students are eligible if and only if they are not freshmen

presents similar difficulties. It may be expressed using '$\equiv$' as

$(x)[Sx \supset (Ex \equiv \sim Fx)]$,

but '$\supset$' must be the main connective. The sentence should not be expressed as '$(x)[(Sx \supset Ex) \equiv \sim Fx]$', for the latter makes an entirely different and much weaker claim.

It is important to notice, also, that words like 'anyone', 'someone', 'everybody', and 'whoever' restrict the individuals being considered to persons. Unless the universe of discourse is explicitly acknowledged to include only persons, then, the translation of sentences containing such words requires use of the open sentence 'x is a person' ('Px', usually). Thus 'Somebody has broken in' should be expressed as '$(\exists x)(Px \ \& \ Bx)$' and 'Whoever broke in left hurriedly' as '$(x)[(Px \ \& \ Bx) \supset Lx]$'.

EXERCISES

I.   Which of the following are true and which false in the universe described earlier in this section? Briefly explain your answers.
1. $(x)[(Gx \ \& \ \sim Fx) \supset Kx]$
2. $(\exists x)[Fx \ \& \ (Gx \equiv Hx)]$
3. $(Fb \vee Gc) \supset (\exists x)(\sim Hx)$
4. $(x)[(Fx \vee Kx) \supset (Hx \ \& \ Gx)]$
5. $(x)(Jx \supset Kx)$
6. $(\exists x)(Jx \ \& \ Kx)$

II. Put each of the following into symbols using the suggested notation.
1. Every woman is fickle. (*Wx-x* is a woman; *Fx-x* is fickle.)
2. Some corporations do not act responsibly. (*Cx-x* is a corporation; *Rx-x* acts responsibly.)
3. If George is allowed to speak, no one will be bored. (*g* - George; *Sx-x* is allowed to speak; *Px-x* is a person; *Bx-x* will be bored.)
4. Horses exist but unicorns do not. (*Hx-x* is a horse; *Ux-x* is a unicorn.)
5. He who laughs last, laughs best. (*Px-x* is a person; *Lx-x* laughs last; *Bx-x* laughs best.)
6. You'll like all our products. (*Lx*-you will like *x*; *Px-x* is one of our products.)
7. If Al or Charlie wins the race, somebody will be disappointed. (*a*-Al; *c*-Charlie; *Wx-x* wins the race; *Px-x* is a person; *Dx-x* will be disappointed.)
8. There are no politicians who are not ambitious. (*Px-x* is a politician; *Ax-x* is ambitious.)
9. Not every politician is ambitious. (*Px-x* is a politician; *Ax-x* is ambitious.)
10. Only a fanatic would advocate violence. (*Fx-x* is a fanatic; *Vx-x* advocates violence.)
11. None but the strong survive. (*Sx-x* is strong; *Vx-x* survives.)
12. Everything enjoyable is either immoral, illegal, or fattening. (*Ex-x* is enjoyable; *Mx-x* is moral; *Lx-x* is legal; *Fx-x* is fattening.)
13. Items purchased at this store are returnable for full refunds if — but only if — they are undamaged and accompanied by the original sales slips. (*Ix-x* is an item purchased at this store; *Rx-x* is returnable for a full refund; *Dx-x* is damaged; *Ax-x* is accompanied by the original sales slips.)
14. Among birds, only ostriches and emus are flightless. (*Bx-x* is a bird; *Ox-x* is an ostrich; *Ex-x* is an emu; *Fx-x* can fly.)
15. There are conservatives who gnash their teeth if they read the *Nation*. (*Cx-x* is a conservative; *Tx-x* gnashes *x*'s teeth; *Nx-x* reads the *Nation*.)
16. The only animals in my house are cats. (*Ax-x* is an animal in my house; *Cx-x* is a cat.)
17. Dolphins and porpoises grin and frolic in the sea. (*Dx-x* is a dolphin; *Px-x* is a porpoise; *Gx-x* grins; *Fx-x* frolics in the sea.)
18. Not only radicals but revolutionaries attended the convention. (*Rx-x* is a radical; *Vx-x* is a revolutionary; *Cx-x* attended the convention.)

19. Any man will be driven to drink who cannot tolerate children. (*Mx-x* is a man; *Dx-x* will be driven to drink; *Tx-x* can tolerate children.)

20. Some politicians will be reelected only if their Mafia connections are not revealed. (*Px-x* is a politician; *Rx-x* will be reelected; *Mx-x*'s Mafia connections are revealed.)

21. No adults will be admitted after 8:00 P.M. unless they are willing to pay double. (*Ax-x* is an adult; *Ex-x* will be admitted after 8:00 P.M.; *Wx-x* is willing to pay double.)

22. Not all adults who are willing to pay double will be admitted after 8:00 P.M. (*Ax-x* is an adult; *Ex-x* will be admitted after 8:00 P.M.; *Wx-x* is willing to pay double.)

23. None think the great unhappy but the great. (*Px-x* is a person; *Tx-x* thinks the great unhappy; *Gx-x* is great.)

24. No Democratic candidate is both a liberal and an advocate of trade restrictions, nor is there any Democratic candidate who is not one of the two. (*Dx-x* is a Democratic candidate; *Lx-x* is a liberal; *Ax-x* advocates trade restrictions.)

25. Except for bluegills, no fish can be hooked without a sinker. (*Bx-x* is a bluegill; *Fx-x* is a fish; *Hx-x* is hooked; *Sx*-a sinker is used to hook *x*.)

*III. Since 'All *J* are *F*' is *true* in the universe specified in this section but 'Some *J* are *F*' is false, the argument

$(x)(Jx \supset Fx)$

$\therefore (\exists x)(Jx \ \& \ Fx)$

is *invalid*. Explain and discuss this rather surprising result, using what you know about the truth-conditions for **A**- and **I**-sentences. Can you suggest a natural additional premise that along with '$(x)(Jx \supset Fx)$' will validly yield '$(\exists x)(Jx \ \& \ Fx)$'?

5.2 Multiple Quantification and Relations

Thus far, our concern has been with general sentences containing a single quantifier, but many sentences are formulable about the universe of the preceding section that contain two or more. Some are straightforwardly truth-functional. For example, the disjunction

(1)   Either everything is *F* or everything is *G*

is *false* in our universe since the component sentences, 'Everything is *F*' and 'Everything is *G*', are both false (*b* is not *F*, and *a* is not *G*). We may express (1) in symbols in either of the following ways:

(1′) $(x)(Fx) \vee (x)(Gx)$

(1″) $(x)(Fx) \vee (y)(Gy)$

Although (1″) is usually preferred, it does not matter whether we use distinct variables for the quantifiers of the component sentences, just as it does not matter whether an open sentence with one variable subject to replacement by a constant is written '*Fx*' or '*Fy*'. For the present we shall continue to use the same variable throughout, though we shall drop the practice when it comes time to consider sentences that *require* the use of distinct variables.

In contrast to (1), the single-quantifier sentence

(2)   Everything is either *F* or *G*

(2′) $(x)(Fx \vee Gx)$

is *true* since each substitution instance of '*Fx* ∨ *Gx*' is true. (Unlike (1′), (2′) is not a truth-functional compound, for though it contains a compound open sentence it does not contain any genuine, "true-or-false" sentences as parts.) Since (1′) is false but (2′) true, it is clear that sentences of the forms

(a) $(x)(\dots x \dots) \vee (x)(--- x ---)$

(b) $(x)(\dots x \dots \vee --- x ---)$

are not equivalent, for we have produced a case where there is a difference in truth-value. Sentences of the form (a) *imply* those of form (b) — an argument with a sentence of the first form as premise and one of the latter as conclusion is valid—but we have just seen that the reverse does not hold; sentences of form (b) do not imply the corresponding ones of form (a).

Similar considerations apply to

(3)   Something is both *F* and *K*

(3′) $(\exists x)(Fx \ \& \ Kx)$

which is simply the **I**-sentence 'Some *F* are *K*' and the sentence

(4)   Something is *F* and something is *K*

(4′) $(\exists x)(Fx) \ \& \ (\exists x)(Kx).$

(3′) is false, since the open sentence '*Fx* & *Kx*' does not have a true substitution instance—'*Fa* & *Ka*', '*Fb* & *Kb*', and '*Fc* & *Kc*' all have false conjuncts. But (4′) is true, since both component sentences '$(\exists x)(Fx)$' and '$(\exists x)(Kx)$' are clearly true. So sentences of the forms

(c) $(\exists x)(\ldots x \ldots \ \& \ \text{---}x\text{---})$

(d) $(\exists x)(\ldots x \ldots) \ \& \ (\exists x)(\text{---}x\text{---})$

are not equivalent, though a sentence of form (c) *implies* one of form (d). But, as the foregoing example shows, an argument with a sentence of form (d) as premise and one of form (c) as conclusion is invalid. This may also be seen by inserting '*x* is a dog' for '...*x*...' and '*x* is a cat' for '---*x*---'.

On the other hand, it can be seen intuitively that the pair of sentences

(5′) $(x)(Fx \ \& \ Gx)$

(6′) $(x)(Fx) \ \& \ (x)(Gx)$

will be true or false under just the same conditions; in fact, of course, both are false in our universe. By the same token, the pair

(7′) $(\exists x)(Fx \lor Gx)$

(8′) $(\exists x)(Fx) \lor (\exists x)(Gx)$

will likewise have the same truth-conditions—both in fact are true in our universe.[7] So we may add to our stock of principles the equivalences:

(e) $(x)(...x... \& ---x---)$
(f) $(x)(...x...) \& (x)(---x---)$

(g) $(\exists x)(...x... \lor ---x---)$
(h) $(\exists x)(...x...) \lor (\exists x)(---x---)$

The foregoing considerations make use, on the intuitive level, of the concepts of equivalence and validity for sentences containing quantifiers. But we have not yet given any precise explanation or analysis of them as regards general sentences. The analysis given in Chapter 2 for truth-functional sentences clearly is not applicable here, since truth-tables cannot be used to treat uncompounded sentences like '$(x)(Fx)$' and '$(\exists x)(Fx \& Gx)$'. Therefore we must eventually provide new accounts of equivalence and validity that will accord with our insights concerning the equivalence of (e) and (f) and of (g) and (h), as well as the validity of inferences from (a) to (b) and from (c) to (d). We shall deal with this in Section 5.3, but first there are further types of sentences demanding our attention.

Let us compare the expressions

(9) $(\exists x)(Fx \equiv Gx)$

(10) $(\exists x)(Fx) \equiv Gx.$

(9) differs in appearance from (10) only in the placement of parentheses, but this difference is very important. For while (9) is a complete sentence and so is true or false, (10) is an *open* sentence having no truth-value. Since 'Gx' occurs outside the parentheses, (10) is a compound having '$(\exists x)(Fx)$' and the open sentence 'Gx' as components. As such, (10) itself is an open sentence, since the 'x' in 'Gx' is subject to replacement by individual constants—though of course neither 'x' in the component '$(\exists x)(Fx)$' is so subject. Thus (10) yields as substitution instances the sentences:

$(\exists x)(Fx) \equiv Ga$ Something is F if and only if a is G
$(\exists x)(Fx) \equiv Gb$ Something is F if and only if b is G
$(\exists x)(Fx) \equiv Gc$ Something is F if and only if c is G.

---

[7]If these equivalences are not apparent to the reader, let him consider whether all substitution instances of '$Fx \& Gx$' could be true without all instances of 'Fx' and of 'Gx' being true, and vice versa; or whether at least one instance of '$Fx \lor Gx$' can be true without at least one instance of 'Fx' or of 'Gx' being true, and vice versa.

But since (9) is a sentence (and so either true or false), *no* occurrence of 'x' in it is subject to replacement by a constant. Rather, the role of 'x' in 'Gx' of (9) is that of a pronoun carrying reference back to the 'x' of the quantifier—*not* of a marker for a position where constants may be substituted. That is, 'x' plays the same role in (9) as 'it' plays in

There is at least one individual such that *it* is *F* if and only if *it* (that same individual) is *G*.

So the parentheses in (9) show that the 'x' of 'Gx' refers back to the 'x' in the quantifier; but the parentheses in (10) show that only the 'x' of 'Fx' refers back, leaving the 'x' of 'Gx' open to substitution. Whereas (9) is formed by attaching '($\exists x$)' to '$Fx \equiv Gx$', (10) is a compound open sentence with '($\exists x$)(Fx)' and the open sentence 'Gx' as components.

We might visualize the situation as follows—the arrows below show that the circled variables carry back reference to the indicated quantifiers, and the underlying brackets show how much of the entire expression is governed by the quantifiers:

(9) $(\exists x)(F \ⓧ \equiv G \ⓧ)$
(10) $(\exists x)(F \ⓧ) \equiv Gx$

We shall describe the portion governed by a quantifier—the portion enclosed in subsequent parentheses—as the **scope** of that quantifier. Thus in (9) the scope of '($\exists x$)' extends throughout the entire remaining expression, but in (10) it extends only to the connective '$\equiv$'. A variable that is *within* the scope of a quantifier containing an occurrence of that same variable always carries back reference, but one *outside* the scope of any quantifier marks a place where a constant may be substituted—it never carries back reference. In this latter case, the entire expression is an open sentence.

We noticed at the outset that it is immaterial what variable is used in an open sentence—we could just as well write 'Fy' as 'Fx'. Thus the open sentence (10) may be rewritten as

(10′) $(\exists x)(Fx) \equiv Gy$

where 'y' is subject to replacement; clearly (10′) admits of the very same substitution instances as (10). But let us now compare

(11) $(\exists x)(Fx \equiv Gy)$.

Here, the parentheses make it clear that the scope of '($\exists x$)' extends over the entire expression, and hence 'y' is within its scope. Even so, 'y' does *not* carry back reference, since the quantifier contains a *different* variable; only the 'x' of 'Fx' serves as a pronoun. Unlike (9), then, but like (10) and (10′), (11) is an open sentence and as substitution instances has:

($\exists x$)($Fx \equiv Ga$)
($\exists x$)($Fx \equiv Gb$)
($\exists x$)($Fx \equiv Gc$).

Like the 'y' in (10′), 'y' here marks a place where an expression designating an individual may be substituted. It is only when 'y' is within the scope of a quantifier like '($\exists y$)' or '(y)' that it carries pronominal cross reference. The importance of quantifier scope may be brought out by noting that the substitution instances of (10′) and (11) are *not* equivalent. For example, if a has F but b does not, and in addition neither a nor b has G, then '($\exists x$)($Fx \equiv Ga$)' will be *true* but '($\exists x$)(Fx) $\equiv Ga$' *false*.

The distinction in role between serving as cross-referring pronoun and marking a place for substitution of constants is sufficiently important to warrant the introduction of some new terminology. We shall describe occurrences of variables in the former role — as well as occurrences within quantifiers themselves — as **bound**, and occurrences in the latter role as **free**. More formally, we have:

DEFINITION   An occurrence of an individual variable is **bound** if and only if it either is part of a quantifier or else lies within the scope of a quantifier containing an occurrence of that same variable. Moreover, an occurrence of an individual variable is **free** if and only if it is not bound, that is, it is neither part of a quantifier nor within the scope of any quantifier containing an occurrence of that same variable.

From these definitions and the foregoing discussion, it follows that:

An open sentence must have an individual variable with at least one free occurrence;

In a true-or-false sentence, all occurrences of all variables (if any) are bound.

In sentences (1′) through (9) all variables occur bound; in (10) 'x' occurs free in 'Gx', and bound elsewhere. In (10′) and (11) 'y' occurs free, but all occurrences of 'x' are bound.

Both '$Fx \equiv Gy$' and '$(\exists x)(Fx \equiv Gy)$' are open sentences, but in the first both 'x' and 'y' occur free whereas in the second only 'y' does. However, in the expression

(12) $(\exists x)(\exists y)(Fx \equiv Gy)$

no variable occurs free, hence it is not an open sentence. Rather, it says that there are individuals x and y such that the first is F if and only if the second is G; this in fact is true in the universe specified in Section 5.1. The parentheses show that the scope of '$(\exists y)$' extends over the remainder of the expression, thus binding the right-hand occurrence of 'y'. We shall leave it understood that when two quantifiers occur juxtaposed as in (12), the one on the left has the other within its scope — parentheses will not be used in such cases. Thus, (12) may be diagrammed:

It should be noted that whereas (1'), (4'), (6'), and (8') can optionally be expressed using distinct variables for the component quantifiers, a sentence like (12) *requires* distinct variables, for otherwise we should not know which variable is bound by which quantifier. Whenever the scope of one quantifier lies within the scope of another, distinct variables are a necessity.

Care must be taken to distinguish open sentences in which more than one variable occurs free — as in '$Fx \equiv Gy$' — from those which, like '$Fx \equiv Gx$' and '$(\exists y)(Fx \equiv Gy)$', contain just one. For while '$Fa \equiv Ga$' and '$Fb \equiv Gb$' are substitution instances of '$Fx \equiv Gx$', the sentence '$Fa \equiv Gb$' is not; different constants cannot replace occurrences of the same variable. On the other hand, the same constant may of course replace distinct variables; thus '$Fx \equiv Gy$' has not only '$Fa \equiv Gb$' but '$Fa \equiv Ga$' and '$Fb \equiv Gb$' as legitimate substitution instances.

Open sentences with more than one free variable have so far all been truth-functionally compound in form, such as '$Fx \equiv Gy$'. There are, however, many that are not compound but are instead *relational*, for example, 'x is larger than y' (in symbols, 'Lxy'). A relational term like 'larger than' may be regarded as another species of general term, but instead of classifying individuals singly, it classifies *pairs* of individuals. Just as a general term like 'philosopher' represents a property of individuals, so a relational term represents a *relation* holding among two or more individuals. The relation represented by 'larger than' holds of pairs of individuals and is thus termed a *dyadic* (or *binary*) relation; it holds, for example, between Boston and Albuquerque respectively, but not between Boston and Chicago or for that matter Albuquerque and Boston.

A *triadic* (or *tertiary*) relation holds among three individuals; 'between' thus holds for the ordered triad Chicago–Milwaukee–Indianapolis, but not the triad Milwaukee–Chicago–Indianapolis. There are also relations that may hold among four, five, and more individuals, but they are less commonly found in ordinary speech. An example of a four-individual relation, however, is provided by

Jim and Angie were married by Preacher Rowe in Deadwood Methodist Church.

Moreover, we saw in Chapter 1 that sentences like

Al and Bill are opponents

cannot sensibly be construed as conjunctions; the above is certainly not a compound of 'Al is an opponent' and 'Bill is an opponent'. Rather, it is a relational sentence, and may be rephrased as

Al is an opponent of Bill,

which goes readily into symbols as '*Oab*'.

As with general terms signifying properties, we shall treat relational terms as supplying the form of the sentences in which they occur. Thus from the singular sentences

Boston is larger than Albuquerque	*Lba*
Boston is larger than Des Moines	*Lbd*
Boston is larger than Chicago	*Lbc*
Albuquerque is larger than Boston	*Lab*

we may abstract the schema '...is larger than---', and as before we shall use capital letters to abbreviate such expressions in our notation. A singular sentence such as '*Lba*' is thus a (true) substitution instance of the open sentence '*Lxy*'; on the other hand, '*Lbx*' is an open sentence with just *one* free variable and has only the first three of the above sentences as substitution instances.

Let us consider some further relational sentences and see how they may be expressed in symbols:

(13) Al struck something

(14) Something struck Al

(15)   Nothing appeals to Bill

(16)   Whatever Alice wants, she gets

(17)   Al hit Bill with a brickbat

(18)   God helps those who help themselves.

Now (13) will be true just in case at least one individual was struck by Al. So using '*Sxy*' for '*x* struck *y*' and '*a*' for 'Al', we get

(13') $(\exists x)(Sax)$

or, in words, 'There is at least one *x* such that *a* struck *x*'. Similarly, (14) is rendered

(14') $(\exists x)(Sxa)$.

Using obvious notation, (15) may be expressed

(15') $(x)(\sim Axb)$.

Now, (16) maintains that for any individual, if Alice wants it she gets it. So using '*a*' for 'Alice', '*Wxy*' for '*x* wants *y*' and '*Gxy*' for '*x* gets *y*', we may express (16) as

(16') $(x)(Wax \supset Gax)$.

In (17) a triadic relation is expressed by '...hit---with——'. Therefore if we use '*Hxyz*', '*a*', '*b*', and '*Bx*' ('*x* is a brickbat'), we obtain

(17') $(\exists x)(Bx \,\&\, Habx)$.

(18) has the force of 'All those who help themselves are such that God helps them', *not* 'All those whom God helps help themselves'. Using '*Hxy*' for '*x* helps *y*' and '*g*' for 'God', we thus derive

(18') $(x)[(Px \,\&\, Hxx) \supset Hgx]$,

where '*Hxx*' conveys the notion of helping oneself.

EXERCISES

I. Which of the following are open sentences? Specify the free occurrences in each case.
1. Fa & $\sim Gx$
2. $Fx \supset (\exists y)(Gy)$
3. $Fa \vee Gb$
4. $(x)[Fx \supset (Gx$ & $\sim Hz)]$
5. $(\exists x)[Fx$ & $(Gx \equiv Hx)]$
6. $(x)(Fx) \supset (Ha \vee \sim Gx)$

II. Consider the open sentence
$(\exists x)(Fx) \supset [Gy \supset (Hxa$ & $\sim Jy)]$.
Which of the following are substitution instances?
1. $(\exists x)(Fx) \supset [Ga \supset (Hba$ & $\sim Jc)]$
2. $(\exists x)(Fx) \supset [Ga \supset (Haa$ & $\sim Ja)]$
3. $(\exists x)(Fx) \supset [Gb \supset (Hca$ & $\sim Jb)]$
4. $(\exists x)(Fa) \supset [Gb \supset (Hca$ & $\sim Jb)]$

Techniques for Translating Complex Sentences

We have seen that the words 'all' and 'any' may often be interchanged without affecting the meaning of the containing expression. Thus, 'All philosophers are wise' and 'Any philosopher is wise' are both expressed as '$(x)(Px \supset Wx)$'. However, there are other contexts where the words play different roles—if you are offered a box of candy and asked to take *any* piece, you surely will commit a social blunder by taking *all* of them. The same contrast holds between 'every' and 'any'. The sentences

I can lick every man in the house
I can lick any man in the house

generally carry the same meaning, and both are rendered as the **A**-sentence '$(x)(Mx \supset Lx)$', where 'Mx' is 'x is a man in the house' and 'Lx' is 'I can lick x'. But note what happens when the important word 'not' is inserted in both:

(19) I cannot lick every man in the house

(20) I cannot lick any man in the house.

Here, 'every' and 'any' operate differently, for (19) says in effect that not every man in the house is one I can lick, that is, there is at least one I cannot lick. But (20) says that *no* man in the house is one I can lick.

Thus (19) is most naturally expressed as an **O**-sentence (and the contradictory of '$(x)(Mx \supset Lx)$'), whereas (20) may be rendered as an **E**-sentence:

(19') $(\exists x)(Mx \;\&\; \sim Lx)$

(20') $(x)(Mx \supset \sim Lx)$.

We must pursue this matter further, for correct translation of many sentences containing 'every', 'any', 'some', and the like can be an intricate task. Let us consider the sentences:

(21)   If everything has mass then something has velocity

(22)   If something has mass then something has velocity

(23)   If anything has mass then something has velocity

(24)   If anything has mass then it has velocity

(25)   If something has mass then it has velocity.

(21) is quite straightforward; we simply put 'Everything has mass' and 'Something has velocity' into symbols and insert a horseshoe between them:

(21') $(x)(Mx) \supset (\exists y)(Vy)$.

Though distinct variables are not necessary, we shall use them for clarity. Likewise, (22) becomes

(22') $(\exists x)(Mx) \supset (\exists y)(Vy)$.

In both (21') and (22'), the scope of the left-hand quantifier extends only to the connective '$\supset$', not over the entire expression. Moreover, (23) is *also* correctly translated as (22'), since the force of 'anything' in (23) is that of 'at least one thing'.

But when we compare (23) with (24), a striking contrast emerges. Since 'it' in (24) carries reference back to the antecedent, we must use the same variable in both antecedent and consequent. But if we now attempt to translate (24) using (23) as our model, we get into trouble. For in

$(\exists x)(Mx) \supset Vx$

we do not have a genuine sentence, but an *open* sentence with a free occurrence of '*x*' in the consequent. Suppose, then, we bind this occurrence by extending the scope of the quantifier across the entire expression, thus getting

$(\exists x)(Mx \supset Vx).$

Now at least we have a sentence that is true or false. However, it is an *incorrect* translation of (24). For it says that there is some individual such that if it has mass then it has velocity, and this lacks the generality of (24). (24) clearly says that *no matter which* individual we pick, it has velocity if it has mass. Therefore, (24) must be translated both using the universal quantifier and with scope extending across the entire expression. Thus, it comes out as the **A**-sentence

(24') $(x)(Mx \supset Vx)$

and means the same as 'Everything (anything) that has mass has velocity.' Moreover, (25) is *also* translated as (24'), since 'something' in such a context clearly carries the same force as 'anything' in (24). Thus 'something' is not always to be translated by means of '$(\exists x)$'. Hence, just as there are English contexts where 'every' and 'any' play the same role, so there are contexts where 'some' and 'any' do. Each of the following is thus translated as (24'):

$\begin{Bmatrix} \text{Everything} \\ \text{Anything} \end{Bmatrix}$ which has mass has velocity

If $\begin{Bmatrix} \text{anything} \\ \text{something} \end{Bmatrix}$ has mass, then it has velocity.

Moreover, (22) and (23) admit of an alternate translation which, like that of (24) and (25), contains a universal quantifier with scope extending over the entire expression. That is, (22) and (23) may be translated not just as (22') but also as

(22'') $(x)[Mx \supset (\exists y)(Vy)].$

(22') and (22'') are equivalent,[8] but two important facts about the latter should be kept in mind: first, the use of distinct variables is essential in

---

[8]If the reader has difficulty seeing the equivalence of (22') and (22''), let him consider, first, whether (22') can be *true* when the open sentence '$Mx \supset (\exists y)(Vy)$' has a *false* substitution instance, and second, whether (22') can be *false* when all substitution instances of '$Mx \supset (\exists y)(Vy)$' are *true*.

(22″), since the scope of the existential quantifier lies within that of the universal; and second, the scope of the universal quantifier *must* extend across the entire expression even though there is no occurrence of '*x*' in the consequent. If it did not, we would have '$(x)(Mx) \supset (\exists y)(Vy)$' — (21′) — which makes the quite different claim that if *everything* has mass, something has velocity. More fully, the contrast between (21′) and (22″) is best appreciated in terms of their English paraphrases:

(21′) *If*, for every *x*, *x* has mass, *then* there is a *y* such that *y* has velocity
(22″) For every *x*, *if x* has mass, *then* there is a *y* such that *y* has velocity.

The equivalence just noted may be stated more generally as follows. Let '…*x*…' represent any open sentence with a free occurrence of '*x*' (for example, '*Fx*', '*Fx* $\supset$ ~*Gx*', '*Fx* v $(\exists y)(Gy)$') and let '*p*' represent any expression whatever as long as it does not contain a free occurrence of '*x*'; then sentences of the following forms are equivalent:

$(\exists x)(…x…) \supset p$
$(x)(…x… \supset p)$,

but these are *not* equivalent to sentences of the form

$(x)(…x…) \supset p$.

Of course the same holds using variables '*y*', '*z*', and so on. Clearly (22′) is derivable from the first schema substituting '*Mx*' for '…*x*…' and '$(\exists y)(Vy)$' for '*p*', and (22″) is derivable from the second using the same substitutions. The nonequivalent (21′) is derivable from the third schema but not, of course, from the second. The restriction that '*p*' contain no free occurrence of the variable that occurs bound in the first quantifier is necessary here, for otherwise that variable would also occur free in the consequent of the first schema but bound in the second. In such a case the equivalence breaks down, since one expression is an open sentence while the other is not.

Relying on what we have just learned, let us tackle more complicated examples:

(26) If all stockholders were in attendance, some of the executives were upset
(27) If any (some) stockholders were in attendance, some of the executives were upset

(28) If any (some) stockholders were in attendance, a board member notified them.

(26) is easily expressed as a conditional with '$(x)(Sx \supset Ax)$' as antecedent and '$(\exists y)(Ey \ \& \ Uy)$' as consequent:

(26′) $(x)(Sx \supset Ax) \supset (\exists y)(Ey \ \& \ Uy)$.

(27) is also a conditional, but with '$(\exists x)(Sx \ \& \ Ax)$' as antecedent:

(27′) $(\exists x)(Sx \ \& \ Ax) \supset (\exists y)(Ey \ \& \ Uy)$.

But since (27′) is of the form '$(\exists x)(\ldots x \ldots) \supset p$' with '$Sx \ \& \ Ax$' substituted for '$\ldots x \ldots$' and '$(\exists y)(Ey \ \& \ Uy)$' for '$p$', we may, according to the equivalence principle just noted, reexpress (27) in the form '$(x)(\ldots x \ldots \supset p)$',

(27″) $(x)[(Sx \ \& \ Ax) \supset (\exists y)(Ey \ \& \ Uy)]$.

Again, the outer brackets are essential, since the scope of '(x)' must extend across the entire expression. Otherwise we would have '$(x)(Sx \ \& \ Ax) \supset (\exists y)(Ey \ \& \ Uy)$', which makes the quite different claim that if *everything whatever* is both a stockholder and in attendance, some executives were upset. Finally, (28) must be handled along the lines of (27″), since 'them' obviously refers back to 'stockholders' in the antecedent. Hence the variable used in the antecedent must also occur in the consequent, and the scope of the universal quantifier must accordingly extend over the entire expression so as to bind that occurrence. Using 'Bx' for 'x is a board member' and 'Nxy' for 'x notified y', we have

(28′) $(x)[(Sx \ \& \ Ax) \supset (\exists y)(By \ \& \ Nyx)]$.

This last example leads us to relational sentences containing more than one quantifier. Beginning with some elementary cases, it should be intuitively clear that

$(x)(y)(Rxy)$
$(y)(x)(Rxy)$

are equivalent; both mean 'Everything bears R to everything'. At best, the only difference between the two expressions would be an insignificant one of phrasing—the latter might more naturally be expressed in the passive voice: 'Everything has R borne to it by everything'. But just as the two English sentences come to the same thing, so do their symbolic counterparts. Moreover, the same holds for the existential quantifier; '$(\exists x)(\exists y)(Rxy)$' and '$(\exists y)(\exists x)(Rxy)$' are equivalent and correspond to 'Something bears R to something'.

On the other hand, order of quantifiers becomes important when one is universal and the other existential, for the following are *not* equivalent:

(29) $(x)(\exists y)(Cxy)$

(30) $(\exists y)(x)(Cxy)$.

To see this, let 'Cxy' mean 'x causes y'; then (29) says

Everything causes something (or other)

or more fully

For any x, there is at least one y such that x causes y.

Notice that this does not say that there is a y which is caused by each and every x. Rather, it makes the weaker claim that for any such x, a y can be found that x causes, but it need not be the *same* y for all the x's—hence the parenthetical phrase 'or other'. However, (30) *does* make the stronger claim—it says that at least one y has every x for a cause, and so corresponds to the English

Something is caused by everything

or, eliminating the passive voice:

Something is such that everything causes it.
There is at least one y such that for any x, x causes y.

Unlike (29), then, (30) will be true only if at least one individual can be found that *all* individuals cause. But though (29) and (30) are not equivalent, it should now be apparent that the stronger claim (30) *implies* (29), for if there is at least one individual that everything causes, then everything causes something (*that* individual if nothing else). The same contrast occurs between:

(30) $(x)(\exists y)(Cyx)$

(31) $(\exists y)(x)(Cyx)$.

(30) corresponds to the English sentences:

Everything is caused by something (or other)
Everything is such that something (or other) causes it
For any x, there is at least one y such that y causes x.

In effect, every individual has a cause, though it need not be the *same* cause for all of them. (31), however, makes the following claim:

Something causes everything
There is a y such that for any x, y causes x.

(31) will be true only if there is an individual (a deity perhaps) that causes everything whatever, but (30) does not have such a stringent requirement for its truth. So they are not equivalent, although (31) *implies* (30).

To amplify a bit, let us consider the more familiar cases:

(32) Every boy likes some girl

(33) Some girl is liked by every boy.

(32) would most naturally be understood as meaning 'some girl or other', while (33) would have the force of 'Some girl is such that she is liked...'. So interpreted, (33) asserts the existence of a girl who is universally liked by the younger male population, whereas (32) makes the more plausible claim that for any boy, a girl can be found whom he likes. (32) says that every boy is such that there is a girl he likes, and (33) says that there is a girl such that every boy likes her. More fully, we have:

(32') For any x, if x is a boy, then there is a y such that y is a girl and x likes y

(33') There is a y such that y is a girl, and for any x if x is a boy then x likes y.

In symbols, then, we derive:

(32″) (x)[Bx ⊃ (∃y)(Gy & Lxy)]

(33″) (∃y)[Gy & (x)(Bx ⊃ Lxy)].

Turning now to further cases of multiple quantification, we consider:

(34) Every suspect who carries a gun will be denied bail

(35) No athlete wins any race unless he trains for it

(36) A rolling stone gathers no moss

(37) Where there's smoke, there's fire.

The technique of *paraphrasing inward* is helpful in translating such sentences. The first objective is to find the *main* quantifier – that with widest scope – and the main connective. Secondly, put both into symbols, leaving the rest of the sentence in English for the moment. Then symbolize as many of the subordinate connectives and open sentences as make themselves apparent. Finally, find the subordinate quantifiers and, in successive steps, put each into symbols along with the remaining English expressions within their scope. (34) provides an elementary example of the technique, for the first step rather obviously yields

(x)(x is a suspect who carries a gun ⊃ x will be denied bail).

However, it is clear that a further quantifier is embedded in the antecedent, for the indefinite article 'a' preceding 'gun' is replaceable with 'any', 'some', and 'at least one'. Thus, using 'Sx' for 'x is a suspect' and 'Bx' for 'x will be denied bail', our next step will be

(x)[(Sx & there is a gun which x carries) ⊃ Bx].

Finally, putting the subordinate quantifier into symbols along with the remaining bit of English within its scope, we have

(34′) (x){[Sx & (∃y)(Gy & Cxy)] ⊃ Bx}

where 'Cxy', of course, means 'x carries y'.

Some readers may have felt inclined to read (34) as saying that *any* suspect who carries *any* gun will be denied bail. Understood this way, the following translation naturally suggests itself:

(34″) $(x)(y)\{[Sx \ \& \ (Gy \ \& \ Cxy)] \supset Bx\}$.

(34″) is also a perfectly acceptable translation and thus is equivalent to (34′). But here the "y-quantifier" is universal and has its scope extending over both the antecedent and consequent of the component open sentence. That they are equivalent may be gleaned from a look at their English counterparts:

For any x, if x is a suspect and there is a gun y which x carries, then x will be denied bail
For any x and y, if x is a suspect and y a gun x carries, then x will be denied bail.

It is essential in (34″) that the y-quantifier have scope extending across the main connective, for the sentence

$(x)\{[Sx \ \& \ (y)(Gy \ \& \ Cxy)] \supset Bx\}$

is definitely *not* equivalent to (34′) and (34″) — it says that if x is a suspect and *everything* whatever is a gun he carries then…, which is hardly the intended meaning.

(35), however, demands a translation along the lines of (34″) rather than (34′). As a first step, we have

$(x)(x$ is an athlete who wins any race $\supset x$ trains for it$)$,

since to say no athlete wins *unless* he trains is to say that every athlete who wins *does* train. Were we now to proceed as we did for (34), we would first get

$(x)[(Ax \ \& \ x$ wins any race$) \supset x$ trains for it$]$,

and since 'it' in the consequent refers back to the antecedent, the same variable must be used in both places, thus giving

$(x)\{[Ax \ \& \ (\exists y)(Ry \ \& \ Wxy)] \supset Txy\}$.

But this cannot be right, since the occurrence of '*y*' in the consequent is *free* — the scope of the *y*-quantifier extends to, but not past, the main connective. To bind this occurrence, the scope of the *y*-quantifier must of necessity extend over both antecedent and consequent; hence a proper translation must take the form of (34″) rather than (34′):

(35′) $(x)(y)\{[Ax \ \& \ (Ry \ \& \ Wxy)] \supset Txy\}$.

Alternately, (35) may be equivalently read as 'For any athlete, any race he wins is one for which he trains'. This most naturally goes into symbols as

(35″) $(x)\{Ax \supset (y)[(Ry \ \& \ Wxy) \supset Txy]\}$.

The translation of (36) is more straightforward. Paraphrasing inward, we first get

$(x)(x \text{ is a rolling stone} \supset x \text{ gathers no moss})$.

Clearly the consequent contains a further quantifier signaled by 'no', and may be more clearly brought into focus by

$(x)(Rx \supset \text{no moss is such that } x \text{ gathers it})$,

where '*Rx*' is '*x* is a rolling stone'. The rather stilted (but clearer) English in the consequent may now be replaced using '*Mx*' and '*Gxy*':

(36′) $(x)[Rx \supset (y)(My \supset \sim Gxy)]$.

(37) is also easily handled by the technique of paraphrasing inward. First, let us find a more perspicuous (though less snappy) English version. Presumably the claim is that any *place* where there is smoke is one at which there is fire as well, so our first step is

$(x)(x \text{ is a place where there is smoke} \supset \text{there is fire at } x)$.

Plainly there is a quantifier in both the antecedent and consequent to be expressed, as is better seen by

$(x)[(Px$ & there is smoke at $x) \supset$ there is fire at $x]$.

So, using 'Axy' for 'x is at y' and other obvious notation, we have

(37') $(x)\{[Px$ & $(\exists y)(Sy$ & $Axy)] \supset (\exists z)(Fz$ & $Azx)\}$.
$$Ayx$$

English idioms often conceal relational terms, and a correct translation often depends on exposing the hidden relational structure. For
example, the sentence

(38) It's a sin to tell a lie

seems unproblematic at first blush, but telling a lie presupposes a teller
and a hearer, neither of whom is explicitly mentioned. Thus (38) must
first be reformulated to bring out the implicit triadic relation:

Anyone who tells a lie to someone (anyone) has sinned.

Thus we would have successively:

$(x)(x$ is a person who tells a lie to someone $\supset x$ has sinned)

$(x)[(Px$ & x tells a lie to someone) $\supset Sx]$

$(x)\{[Px$ & $(\exists y)(Ly$ & x tells y to someone)] $\supset Sx\}$

(38') $(x)\{\{Px$ & $(\exists y)[Ly$ & $(\exists z)(Pz$ & $Txyz)]\} \supset Sx\}$

where '$Txyz$' means 'x tells y to z'.
Sentences containing words like 'always', 'sometimes', 'never',
'whenever' usually have temporal significance and hence require an open
sentence like 'x is a time'. For example,

(39) Barry always goes into a rage when George delivers a speech

means roughly 'Anytime George delivers a speech is one at which Barry
goes into a rage'. So using '$Dxyz$' for 'x delivers y at z', 'Rxy' for 'x goes

into a rage at y', and other obvious notation, (39) becomes

(39′) $(x)\{[Tx \ \& \ (\exists y)(Sy \ \& \ Dgyx)] \supset Rbx\}$.

On the other hand, there are contexts where 'always' and the like do not carry temporal significance, for example:

(40)   The cube of a positive integer is always a positive integer.

This really says no more than that every cube of a positive integer is itself a positive integer; hence no open sentence 'x is a time' is needed:

(40′) $(x)(y)[(Ix \ \& \ y = x^3) \supset Iy]$.

The paraphrasing-inward technique is shown to its best advantage in translating highly complex sentences with several quantifiers:

(41)   Any politician who votes against every reform bill will be defeated by a candidate who offers something to every minority group.

Digging out the main quantifier and connective, we have

$(x)(x$ is a politician who votes against every reform bill $\supset x$ will be defeated by a candidate who offers something to every minority group).

In the consequent, we have presumably the claim that *any* candidate who offers something to every minority group will defeat x; so using obvious notation, our next step is:

$(x)\{(Px \ \& \ x$ votes against every reform bill$) \supset (y)[(Cy \ \& \ y$ offers something to every minority group$) \supset Dyx]\}$.

Next, we put the remainder of the antecedent into symbols; it says that every reform bill is such that x votes against it.

$(x)\{[Px \ \& \ (z)(Rz \supset Vxz)] \supset (y)[(Cy \ \& \ y$ offers something to every minority group$) \supset Dyx]\}$.

Now the remaining English clause in the consequent contains an ambiguity. It might mean either of the following:

Something is such that *y* offers it to every minority group
Every minority group is such that *y* offers it something (or other).

Clearly the latter is the intended meaning; the sentence does not claim that the candidate will offer the same plum to minorities with quite different interests and needs. So where '*Oxyz*' means '*x* offers *y* to *z*', we finally have (resorting now to the extra variable '*x''*'):

$$(x)\{[Px \ \& \ (z)(Rz \supset Vxz)] \supset (y)\{\{Cy \ \& \ (w)[Mw \supset (\exists x')(Oyx'w)]\} \supset Dyx\}\}.$$

EXERCISES

I.   Put each of the following into symbols, using the suggested notation.

1.   If every bureau member is a spy, then no one in the government is beyond suspicion. (*Bx-x* is a bureau member; *Sx-x* is a spy; *Px-x* is a person; *Gx-x* is in the government; *Yx-x* is beyond suspicion.)

2.   If any bureau member is a spy, then no one in the government is beyond suspicion. (Same notation as (1).)

3.   If any bureau member is a spy, he will not be arrested by the F.B.I. unless someone turns state's evidence. (*Bx-x* is a bureau member; *Sx-x* is a spy; *Axy-x* will arrest *y*; *f*-the F.B.I.; *Px-x* is a person; *Ex-x* turns state's evidence.)

4.   If somebody wants to run for the Presidency, he'd better have a well-financed campaign. (*Px-x* is a person; *Wx-x* wants to run for the Presidency; *Bx-x* had better have a well-financed campaign.)

5.   Every cloud has a silver lining. (*Cx-x* is a cloud; *Sx-x* is a silver lining; *Hxy-x* has *y*.)

6.   Some senators can outtalk every woman. (*Sx-x* is a senator; *Wx-x* is a woman; *Oxy-x* can outtalk *y*.)

7.   Anyone who listens to a prohibitionist is either mad or naive. (*Px-x* is a person; *Bx-x* is a prohibitionist; *Lxy-x* listens to *y*; *Mx-x* is mad; *Nx-x* is naive.)

8.   Alma is jealous of any girl liked by Burt. (*a*-Alma; *Jxy-x* is jealous of *y*; *Gx-x* is a girl; *Lxy-x* likes *y*; *b*-Burt.)

9.   Everybody likes something, but nobody likes everything. (*Px-x* is a person; *Lxy-x* likes *y*.)

10.   An opportunist will do anything for a buck. (*Ox-x* is an opportunist; *Dxyz-x* will do *y* for *z*; *Bx-x* is a buck.)

11. Valid arguments are derivable only from valid argument forms. (Vx-x is a valid argument; Dxy-x is derivable from y; Fx-x is a valid argument form.)

12. Only valid arguments are derivable from valid argument forms. (Same notation as (11).)

13. Nothing ventured, nothing gained. (Vxy-x ventures y; Gxy-x gains y.)

14. Anything not at rest perseveres in motion unless acted upon by an external force. (Rx-x is at rest; Px-x perseveres in motion; Axy-x acts upon y; Fx-x is an external force.)

15. Anyone who keeps everything he earns for himself is neither generous nor prudent. (Px-x is a person; $Kxyz$-x keeps y for z; Gx-x is generous; Rx-x is prudent; Exy-x earns y.)

16. Delilah had a ring on every finger and had a finger in every pie. (d-Delilah; Oxy-x is on y; Rx-x is a ring; Fxy-x is a finger of y; Px-x is a pie; Ixy-x is in y.)

17. Any man who hates children and dogs has W.C. Fields for a friend. (Mx-x is a man; Hxy-x hates y; Cx-x is a child; Dx-x is a dog; f-Fields; Fxy-x has y for a friend.)

18. Hell hath no fury like a woman scorned. (h-Hell; $Hxyz$-x hath y like z; Fx-x is a fury; Wx-x is a woman; Sxy-x scorns y.)

19. A fool and his money are soon parted. (Fx-x is a fool; Mx-x is money; Bxy-x belongs to y; Pxy-x is soon parted from y.)

20. Any watch sold to anyone by Crumley is a stolen item. (Wx-x is a watch; $Sxyz$-x sells y to z; c-Crumley; Px-x is a person; $Lxyz$-x stole y from z.)

21. There is a politician who is despised by all citizens who despise any politician at all. (Px-x is a politician; Dxy-x despises y; Cx-x is a citizen.)

22. A watched pot never boils. (Px-x is a pot; $Wxyz$-x watches y at z; Bxy-x boils at y; Tx-x is a time.)

23. There is a prime number between 14 and 18 greater than any number that is a square root of a number less than 300. (Px-x is prime; Nx-x is a number; f-14; e-18; $x > y$-x is greater than y; $x = \sqrt{y}$-x is a square root of y; $x < y$-x is less than t; $Bxyz$-x is between y and z; a-300.) By the way, is (23) true?

24. A politician who appeases every pressure group evades every issue with a platitude. (Px-x is a politician; Axy-x appeases y; Gx-x is a pressure group; Ix-x is an issue; $Exyz$-x evades y with z; Tx-x is a platitude.)

25. If a politician caters only to businessmen, then unless he is influential and maintains a strong organization none of them will return the compliment. (Px-x is a politician; Axy-x caters

to y; Bx-x is a businessman; Ix-x is influential; Ox-x is a strong organization; Mxy-x maintains y; Rx-x returns the compliment.)

II. Let us add to our notation some standard mathematical symbols, and suppose our universe of discourse to be restricted to the series of positive integers: 1, 2, 3, *ad infinitum*. Which of the following are true and which false? Explain briefly in each case.

1. $(x)(\exists y)(y > x)$
2. $(x)(\exists y)(x > y)$
3. $(\exists y)(x)(y > x)$
4. $(\exists y)(x)(y \neq x \supset y > x)$
5. $(\exists x)(y)(y \neq x \supset y > x)$

In each of the following '$Bxyz$' means 'x is between y and z'.

6. $(\exists x)(\exists y)(\exists z)(Bxyz)$
7. $(x)(\exists y)(\exists z)(Bxyz)$
8. $(y)(\exists x)(\exists z)(y \neq 1 \supset Bxyz)$
9. $(\exists y)(x)(\exists z)(x \neq y \supset Bxyz)$
10. $(x)(y)(z)[Bxyz \equiv (z < x \ \& \ x < y)]$

III. The following remark, often attributed to Lincoln, is ambiguous:
   You can fool some of the people all of the time.
   State first in English and then in symbols the two different meanings. Be sure to specify your notation.

5.3 Interpretations and Invalidity

In the course of our discussion we have discovered several equivalence principles and valid argument forms. While we did not *prove* equivalence or validity in these cases, they were sufficiently obvious to be grasped intuitively. We know, for example, that each of the following sentence pairs are equivalent:

$\sim(\exists x)(Fx)$	and	$(x)(\sim Fx)$
$(\exists x)(\sim Fx)$	and	$\sim(x)(Fx)$
$(x)(Fx \;\&\; Gx)$	and	$(x)(Fx) \;\&\; (y)(Gy)$
$(\exists x)(Fx \lor Gx)$	and	$(\exists x)(Fx) \lor (\exists y)(Gy)$
$(\exists x)(Fx) \supset (\exists y)(Gy)$	and	$(x)[Fx \supset (\exists y)(Gy)]$

Moreover, we discussed some elementary cases of validity. Each of the following arguments, we know, is valid, although when the premise and conclusion are interchanged the resulting argument is invalid in each case.

$(\exists x)(Fx \;\&\; Gx)$	$(x)(Fx) \lor (y)(Gy)$	$(\exists y)(x)(Cyx)$
$\therefore (\exists x)(Fx) \;\&\; (\exists y)(Gy)$	$\therefore (x)(Fx \lor Gx)$	$\therefore (x)(\exists y)(Cyx)$

Further, it is natural to suppose that there are sentences of quantificational logic that are analogues of the tautologies of truth-functional logic, that is, sentences that are true under all conditions (though we have yet to say just what this phrase comes to for sentences containing quantifiers). Thus, consider the sentence

(1) $(x)(Fx \lor \sim Fx)$.

Clearly this will be true no matter what property we take the predicate letter to represent and no matter what individuals make up our universe of discourse. Thus (1) contrasts sharply with, say, '$(x)(Fx \supset Gx)$', which comes out *true* when 'F' is taken to represent the property of being a man, 'G' represents the property of being mortal, and our universe of discourse is the set of human beings. But it comes out *false* for that same universe when 'F' represents being mortal and 'G' being a man (if 'a' designates a specific woman, then '$Fa \supset Ga$' is false in this latter case). We shall call sentences that, like (1), are true under all conditions **logical truths**. Further examples are:

(2) $(x)[(y)(Fy) \supset Fx]$

(3) $(x)[(\exists y)(\sim Gxy) \supset \sim(z)(Gxz)]$

(4) $\sim (\exists x)(Fx) \equiv (y)(\sim Fy)$

(5) $(\exists x)(Fx \mathbin{\&} Gx) \supset [(\exists y)(Fy) \mathbin{\&} (\exists z)(Gz)]$.

Again, no matter what universe of discourse is assumed nor which properties the predicate letters represent, (2) through (5) come out true.[9]

Now two main tasks lie before us. First, we must precisely *define* the concepts of logical truth, validity, and equivalence for quantificational logic, just as we gave parallel definitions in Chapter 2 of similar notions for truth-functional logic. Second, we must develop a method for *establishing* logical truth, validity, and equivalence — one that will serve the same purpose as the methods of deduction and truth-tables served for truth-functional sentences. The second job is left for Chapter 6; the first is handled in the present section.

These are quite separate tasks. A proper characterization of validity does not automatically yield a method for establishing it. However, as we shall soon see, it easily yields methods for establishing *invalidity* (as well as *nonequivalence*, and so on). This is important, for the method developed in Chapter 6 is an extension of the method of deduction introduced in Chapter 3, and it is therefore a method for establishing validity, equivalence, and logical truth *only* — not for their respective opposites.[10] Unfortunately, in quantificational logic with relations and multiple quantifiers, there is no *mechanical* routine analogous to the truth-tables of sentential logic for determining *both* validity and invalidity, equivalence and nonequivalence, and so on. This is not to say merely that no such method has thus far been discovered; rather, it has been demonstrated that there is no such method to be found.[11] Methods such as those developed in Chapter 6, then, are all we have for showing validity, equivalence, and logical truth, and the decidedly nonmechanical methods to be presented here must suffice for their opposite numbers.

---

[9]It should be plain from the above that what has been called a "sentence" of our symbolic notation, such as '$(x)(Fx \supset Gx)$', differs in an important respect from a sentence of English like 'All men are mortal'. For while the latter has a fixed meaning, the former does not. The letters 'F' and 'G' can represent *different* properties in different contexts, as the foregoing example shows, so that in *one* context '$(x)(Fx \supset Gx)$' means in our notation what is (permanently) meant in English by 'All men are mortal', and in another what is means by 'All mortals are men'. Whether a sentence of our notation is true or false in a given context depends upon which properties the predicate letters have been taken to represent — only then do we know which English sentence it abbreviates in that context. In calling a sentence like (1) a logical truth, we mean that every English sentence it can abbreviate is true in any universe of discourse, for example, 'Everything is either red or not red'.

[10]More precisely, the method allows one to establish a sentence as a logical truth by establishing it as a theorem (deducible from no premises) and to establish two sentences as equivalent by showing them interdeducible with one another.

[11]See Alonzo Church, "A Note on the Entscheidungsproblem," *Journal of Symbolic Logic*, vol. 1 (1936), pp. 40–41.

We can perhaps best develop the desired accounts of validity, equivalence, and logical truth by first considering how invalidity of quantificational arguments may be established. Let us return to an invalid argument noted in the preceding section:

(I) $(x)(\exists y)(Cyx)$
   $\therefore (\exists y)(x)(Cyx)$.

We interpreted 'Cyx' as 'y causes x', and on this interpretation the argument becomes:

(A) Everything has a cause
   Therefore something causes everything.

But of course there are many arguments in English that are of exactly the same form and hence may also be represented in our notation by (I). For example, if 'Cyx' is now interpreted as 'y creates x', we have:

(B) Every being was created by some being
   Therefore some being created every being.

Now if we restrict our universe of discourse to persons — thus eliminating the need for 'Px' — and interpret 'Cyx' as 'y is an acquaintance of x', (I) symbolizes the argument:

(C) Everyone has someone for an acquaintance
   Therefore someone is an acquaintance of everyone.

Now suppose someone stubbornly refuses to acknowledge the invalidity of arguments (A) and (B). We might then offer (C) to show their invalidity, for as their mutual translation by means of (I) reveals, (C) is of the same form as the other two and it is far more obvious that the premise is true and the conclusion false. Surely everyone is such that someone or other is an acquaintance of his, but it is, to say the least, implausible that someone is an acquaintance of every single person. This technique is usually called *refutation by logical analogy* and is commonly employed in practical debate; one attempts to refute his opponent's argument by citing an argument with the same form having obviously true premise(s) and false conclusion. Since the latter argument is invalid, then of course any argument of exactly similar form is invalid as well.

Suppose, however, that our antagonist deftly points out that as 'acquaintance' is ordinarily understood, newborn babies can hardly be said to have acquaintances at all. They therefore provide the needed counterexample for showing the premise of (C) to be false. We must then respond by producing an argument whose invalidity is more blatantly obvious even than (C), and it is not hard to find. Let us this time restrict the universe of discourse to the natural numbers, 0, 1, 2, 3,... and interpret 'Cyx' as 'y is greater than x'. Then (I) translates the argument:

(D) For every number there is a number greater than it
Therefore some number is greater than every number.

Clearly the premise is true, for no matter what number, n, we pick, there is another — its successor $n + 1$ — that is greater. But the conclusion is false — no number is greater than *all* numbers, since among other things no number can be greater than *itself*.

We can achieve the same end by considering a much smaller universe of discourse, one with just two individuals: Mount Everest and Mount Rushmore. Interpreting 'Cyx' as 'y differs in size from x', (I) now symbolizes:

(E) Everything differs in size from something or other
Therefore something differs in size from everything.

Since Everest and Rushmore do in fact differ, the premise is true, and since neither differs in size from itself, the conclusion is false.

Let us now take stock of what we have done here. We showed (A) invalid by logical analogy with arguments (D) and (E), for they have the same form as (A) and in addition an obviously true premise with obviously false conclusion. More importantly, we showed argument (I) of our symbolic notation to be invalid by first specifying a universe of discourse and then assigning a dyadic relation to the letter 'C' such that when (I) was interpreted in this way, we obtained a true premise and false conclusion. Each of (A) through (E) may be regarded as the result of *interpreting* (I) in different ways; hence to show (I) invalid, we need only find at least one interpretation yielding a true premise and false conclusion. To show an argument *in English* invalid, we first put it in symbols. It then remains to produce an interpretation — in effect, another English argument of that form — wherein all premises are clearly true and the conclusion is just as clearly false.

This notion of "interpreting" one or more sentences of our notation is essential for explicating the concepts of validity, equivalence, and logical truth as they pertain to quantificational logic. Let $\mathcal{T}$ be any sentence of our notation. To give an **interpretation** of $\mathcal{T}$, one must first *specify a universe of discourse over which the variables of $\mathcal{T}$ are to range* (if any individual constant occurs in $\mathcal{T}$ it is necessary also to specify which individual of that universe the constant designates), and second, *assign a property or relation* (as the case may be) *to each distinct predicate letter occurring in $\mathcal{T}$*.[12]

Assigning a property or relation to a predicate letter is described as *interpreting the letter*; so to *interpret a sentence*, we must specify a universe of discourse and also interpret each predicate letter in the sentence.

We may thus give an interpretation of the sentence

$(x)[Px \supset (\exists y)(Lxy)]$

as follows:

universe of discourse: 1, 2, 3, . . .
interpret '*Px*' as '*x* is a prime'
~~universe of discourse: 1, 2, 3, . . .~~
interpret Lxy as 'y is less than x'

Clearly the sentence comes out true on this interpretation; for each positive integer, *x*, if it is a prime number, then there is in this universe a *y* such

---

[12]We shall for convenience extend the use of 'predicate letter' to include those letters that represent *relations* in a given context. We are now in a position to define recursively a quantificational sentence of our notation:
  i. Each predicate letter followed by a sequence of constants is a sentence (for example, '*Fa*', '*Bab*', '*Ljct*');
  ii. The result of prefixing '~' to a sentence is a sentence;
  iii. The result of inserting '&', 'v', '⊃', or '≡' between two sentences, enclosing the result in parentheses, is a sentence;
  iv. The result of replacing one or more constants in a sentence with some variable μ ('*x*', '*y*', '*z*', and so on) not already occurring in it, enclosing the result in parentheses unless it is already so enclosed, and prefixing either $(\exists\mu)$ or (μ) to the whole, is a sentence;
nothing is a sentence unless it can be shown to be so on the basis of the above clauses.

In practice we sometimes replace parentheses with brackets and braces in the interest of readability, and often omit parentheses when doing so causes no ambiguity—for example, we write '$(x)(y)(Fxy)$' instead of '$(x)((y)(Fxy))$'. Some authors prefer adding a "clause 0" at the beginning:

Each sentence letter ('*A*', '*B*', '*C*', and so on) is a sentence.

If this addition is made, the semantics presented here must be expanded slightly.

that *x* is greater than it (the number 1 will always do for '*y*', since it is not prime). Another interpretation is:

universe of discourse: human beings
interpret '*Px*' as '*x* is a man'
interpret '*Lxy*' as '*x* is father of *y*'.

Here the sentence comes out false, since not every man is a father.

Now, an argument consists of two or more sentences one of which is taken to follow from the others. So we need to introduce the notion of interpreting *several* sentences jointly. To this end, let us reconsider the universe cited in Section 5.1:

	F	G	H	J	K
a	+	−	+	−	−
b	−	+	+	−	+
c	+	+	+	−	−

This chart in effect gives us an interpretation of the sentences of our notation introduced in Sections 5.1 and 5.2. However, it was not specified there just which properties the letters were representing nor which three individuals comprised our universe of discourse. We now make this explicit:

universe of discourse: the integers 7, 6, 2 designated respectively by '*a*', '*b*',
   and '*c*'
interpret '*Fx*' as '*x* is prime'
interpret '*Gx*' as '*x* is even'
interpret '*Hx*' as '*x* is positive'
interpret '*Jx*' as '*x* is a square root of a negative integer'
interpret '*Kx*' as '*x* is divisible by 3'.

The foregoing provides an interpretation not only of the sentences cited earlier about this world, but also of the sentences comprising the *invalid* argument:

(6) $(x)(Hx) \lor (y)(Jy)$

(7) $(\exists x)(\sim Jx) \,\&\, (\exists y)(Gy)$

(8) $(\exists x)(Kx \lor Gx)$

(9) $\therefore Fa \supset (\exists x)(Hx \,\&\, Kx)$

For example, (6) on this interpretation says that either everything (in the specified universe) is positive or else everything is a square root of a negative integer; (9) says that if 7 is a prime number then something is both positive and divisible by 3. Both are true on the above interpretation, as are (7) and (8); hence if this argument is in fact invalid, the above interpretation will not show it so, since the conclusion is not false. However, the following interpretation will do the job:

universe of discourse: Lincoln, Washington, and Franklin D. Roosevelt, designated respectively by 'a', 'b', and 'c'
interpret 'Fx' as 'x is a Republican'
interpret 'Gx' as 'x is a Democrat'
interpret 'Hx' as 'x is a former U.S. President'
interpret 'Jx' as 'x is a former commanding general'
interpret 'Kx' as 'x is a woman'.

This new interpretation gives us:

	F	G	H	J	K
a	+	−	+	−	−
b	−	−	+	+	−
c	−	+	+	−	−

And from this it is easily seen that the conclusion is false, since we now have a true antecedent and false consequent. Moreover, all premises come out true on this interpretation, so the argument has been shown invalid. Just as an invalid truth-functional argument must have at least one row with **T**'s for all premises and **F** for the conclusion, so an invalid quantificational argument must have at least one interpretation of its sentences yielding all true premises and false conclusion. More fully:

A quantificational argument is **invalid** if and only if there is at least one universe of discourse and at least one way of interpreting all predicate letters occurring in its sentences such that all premises are rendered true and the conclusion false.

The foregoing suggests that we should simply define 'valid' in terms of there being no interpretation in which all premises are true and the conclusion is false. This is in fact what we shall do, but first a qualification must be made. By 'no interpretation' we must mean 'no *non-empty* universe of discourse...'. That is, the case of the empty universe — one with no individuals at all — must be excluded from consideration. This universe exhibits counterintuitive characteristics; because of its unique and curious nature it is best that it be excluded from consideration. For example, the plainly valid argument

$(x)(Fx)$
$\therefore (\exists x)(Fx)$

comes ~~to very little, since the universes presupposed in actual contexts of~~ *out invalid in the empty universe (though of course valid in all non-* empty ones). For its conclusion is false in that universe since it asserts the existence of at least one individual; but its premise is *true*, since — as a little reflection shows — it is equivalent to '$\sim(\exists x)(\sim Fx)$', which *denies* the existence of something that is not F. Indeed, a great many intuitively valid arguments have true premises and false conclusion when interpreted with respect to the empty universe. Hence we shall restrict ourselves to universes with at least one individual. This restriction in practice comes to very little, since the universes presupposed in actual contexts of argument generally are known to be non-empty.

Our definitions proceed as follows:

DEFINITION   A quantificational sentence is a **logical truth** if and only if it is true on every interpretation.

DEFINITION   A quantificational sentence is **satisfiable** if and only if it is true on at least one interpretation.

DEFINITION   A quantificational sentence is a **logical falsehood** if and only if it is false on every interpretation.

Thus, '$(x)(Fx \supset Fx)$' is a logical truth, '$(x)(Fx \supset Gx)$' is satisfiable, and '$(\exists x)(Fx \& \sim Fx)$' is a logical falsehood.

DEFINITION   A quantificational argument with premises $\mathscr{P}_1, \mathscr{P}_2, \ldots, \mathscr{P}_n$ and conclusion $\mathscr{C}$ is **valid** if and only if there is no interpretation upon which all of $\mathscr{P}_1, \mathscr{P}_2, \ldots, \mathscr{P}_n$ are true and $\mathscr{C}$ false.

DEFINITION   Two quantificational sentences are **equivalent** if and only if there is no interpretation upon which one is true and the other false.

Validity may also be defined in terms of the corresponding conditional being a logical truth, and equivalence in terms of the logical truth of the corresponding biconditional. To show two sentences *not* to be equivalent, it suffices to exhibit any interpretation upon which they differ in truth-value.

EXERCISES

I.   Find an interpretation showing the *converse* of the logical truth (5) above to be satisfiable and one showing it *not* to be a logical truth.

II. For each of the following arguments, find an interpretation that shows it invalid.
   1. $(x)(Ax \supset Bx)$
        $(\exists x)(Cx \;\&\; {\sim}Bx)$
        $\therefore (\exists x)(Ax \;\&\; {\sim}Cx)$
   2. Rab
        $\therefore (\exists x)(Rxx)$
   3. $(x)[Fx \supset (\exists y)(Gy \;\&\; {\sim}Rxy)]$
        $\therefore (\exists y)[Gy \;\&\; (x)(Fx \supset {\sim}Rxy)]$
   4. $(x)(Fx) \supset (\exists y)(Gy)$
        $\therefore (x)[Fx \supset (\exists y)(Gy)]$
   5. $(x)(\exists y)(Fx \equiv {\sim}Fy)$
        $\therefore (\exists y)(Fy \equiv {\sim}Fy)$
   6. $(x)(Fx \supset Gx)$
        $(\exists x)(Fx \;\&\; Hx)$
        $(\exists x)(Gx \;\&\; {\sim}Fx)$
        $\therefore (\exists x)(Gx \;\&\; {\sim}Hx)$

III. Show each pair to be nonequivalent:
   1. $(x)(Fx \supset Gx)$ and $(x)(Fx) \supset (y)(Gy)$
   2. $(x)[(y)(Fxy) \supset Gx]$ and $(x)(y)(Fxy \supset Gx)$

Often, by deliberately focusing on universes containing a very small number of individuals, invalidity can be established quickly and in more systematic fashion. Let us examine the argument:

(II) $(x)(Fx \supset Gx)$
   $(\exists x)(Gx \;\&\; Hx)$
  $\therefore (\exists x)(Fx \;\&\; Hx)$.

Now consider a universe of discourse with just one individual, *a*. Then in *that* universe, the first premise will be true under the same conditions as '*Fa* $\supset$ *Ga*', since the latter is the only substitution instance of '*Fx* $\supset$ *Gx*'. By the same token, the second premise will be true in this universe just in case '*Ga* & *Ha*' is true, and the conclusion just in case '*Fa* & *Ha*' is true. By assigning truth-values to the simple components '*Fa*', '*Ga*', and '*Ha*', we can make the premises true and conclusion false of the truth-functional argument:

(III) $Fa \supset Ga$
    $Ga \;\&\; Ha$
  $\therefore Fa \;\&\; Ha$.

But if the assignment shows *this* argument invalid, then it shows (II) invalid as well. Since the premises and conclusion of (II) have the same truth-conditions for a one-individual universe as the corresponding premises and conclusion of (III), we have in effect shown that in any universe with just one individual there is an interpretation of the predicate letters of (II) yielding true premises and false conclusion. In the present case, such a truth-value assignment is easily come by:

Fa	*Ga*	*Ha*
F	**T**	**T**

Now consider again the invalid inference:

(IV) (∃*x*)(*Fx*)
      (∃*x*)(*Gx*)
   ∴ (∃*x*)(*Fx* & *Gx*).

Here, for a one-individual universe we would have the corresponding truth-functional argument:

Fa
Ga
∴ *Fa* & *Ga*

which is valid by the rule of **Conj**. Yet we know that there is *some* universe of discourse in which it will have true premises and false conclusion. So let us try a universe with two individuals, *a* and *b*. Under what conditions will the first premise of (IV) be true in such a universe? Since '*Fx*' has two substitution instances, '*Fa*' and '*Fb*', clearly '(∃*x*)(*Fx*)' will be true just in case at least one of them is true, that is, just in case the disjunction '*Fa* ∨ *Fb*' is true. Similar considerations hold for the second premise and conclusion; the latter, for example, has the same truth-conditions in a two-individual universe as '(*Fa* & *Ga*) ∨ (*Fb* & *Gb*)'. So for such a universe, the corresponding truth-functional argument would be:

Fa ∨ *Fb*
Ga ∨ *Gb*
∴ (*Fa* & *Ga*) ∨ (*Fb* & *Gb*).

We may now produce a truth-value assignment yielding true premises and false conclusion, thus showing this argument — and hence (IV) — invalid:

Fa	Ga	Fb	Gb
T	F	F	T

Generally, for a universe with n individuals, the sentence '$(\exists x)(Fx)$' will have the same truth-conditions as the disjunction

Fa_1 v Fa_2 v ... v Fa_n.[13]

Similarly, a universally quantified sentence such as '$(x)(Fx)$' will be true just in case *all* substitution instances of 'Fx' are true and therefore just in case the conjunction of those instances is true. So for an n-individual universe, '$(x)(Fx)$' will be true under the same conditions as

Fa_1 & Fa_2 & ... & Fa_n.

We shall call such disjunctions and conjunctions **truth-functional expansions** of the corresponding existentially and universally quantified sentences. Further examples of such expansions for a three-individual universe are:

$(x)(Fx$ v $Gx)$:	$(Fa$ v $Ga)$ & $(Fb$ v $Gb)$ & $(Fc$ v $Gc)$
$(x)(Fx)$ v $(y)(Gy)$:	$(Fa$ & Fb & $Fc)$ v $(Ga$ & Gb & $Gc)$
$(\exists x)(Fx)$ & $(\exists y)(Gy)$:	$(Fa$ v Fb v $Fc)$ & $(Ga$ v Gb v $Gc)$

Multiply-quantified sentences with relational terms are a bit more complicated, since each quantifier must be broken down separately. Consider again

$(x)(\exists y)(Cyx)$.

Breaking down the main quantifier first, we derive for a two-individual universe

$(\exists y)(Cya)$ & $(\exists y)(Cyb)$.

---

[13]For brevity, we shall here drop parentheses in extended disjunctions and conjunctions.

Breaking down the existential quantifiers then gives us

(*Caa* v *Cba*) & (*Cab* v *Cbb*).

By considering several interpretations, we saw earlier in this section that the argument

(I) $(x)(\exists y)(Cyx)$
  $\therefore (\exists y)(x)(Cyx)$

is invalid, and now we can show its invalidity by first obtaining the truth-functional expansion for the conclusion. Thus, '$(\exists y)(x)(Cyx)$' becomes successively

$(x)(Cax)$ v $(x)(Cbx)$
(*Caa* & *Cab*) v (*Cba* & *Cbb*).

We then show that the argument

(*Caa* v *Cba*) & (*Cab* v *Cbb*)
$\therefore$ (*Caa* & *Cab*) v (*Cba* & *Cbb*)

is truth-functionally invalid. The following assignment of truth-values renders the premise true and the conclusion false:

Caa	Cab	Cba	Cbb
F	T	T	F

If *a* is Mt. Everest and *b* is Mt. Rushmore, the interpretation presented earlier leads directly to this truth-value assignment.

However, not every invalid quantificational argument can be *proved* invalid through the use of truth-functional expansions. Consider, for example,

$(x)(y)(z)[(Ryx \ \& \ Rzy) \supset Rzx]$
$(x)(\exists y)(Ryx)$
$\therefore (\exists x)(Rxx).$

Though this argument is invalid (let '*Ryx*' be '*y* is greater than *x*' and the universe of discourse the series of positive integers), the reader will look in vain for a truth-value assignment that will prove it so. The reason is that it exhibits true premises and false conclusion only for a universe with *infinitely* many individuals. It would of course be impossible to actually write out truth-functional expansions—much less find truth-value assignments. Invalid arguments containing no *relational* letters, however, are unproblematic in this respect.

Moreover, for quantificational arguments containing no relational expressions, the following result[14] holds: where the number of predicate letters is *n*, if there is no truth-value assignment yielding true premises and false conclusion for a universe of 2^n individuals, then the argument is valid—no assignment in *any* non-empty universe will yield such a combination. So, in theory at least, one could mechanically test such arguments for validity *and* invalidity by going directly to the 2^n-individual case. In practice, however, this is usually an unwise strategy. Most invalid arguments may be so established using a much smaller universe, and even for a relatively simple argument with only four predicate letters, the strategy would call for a universe of no less than 16 individuals—along with the correspondingly huge truth-functional expansions.

The 2^n-individual strategy nonetheless constitutes a mechanical routine for determining validity, equivalence, and logical truth as regards non-relational sentences and the arguments in which they occur. However, as noted earlier, there is no such routine for quantificational logic with relations and multiple quantifiers.

EXERCISES

I.   Go back and do the examples in (II) of the preceding exercise set using truth-functional expansions. Then try this one:

$(x)[Fx \lor (Gx \ \& \ Hx)]$

$(\exists x)[Ex \ \& \ (Gx \ \& \ Hx)]$

$(\exists x)[Ex \ \& \ (\sim Gx \lor \sim Hx)]$

$\therefore (x)(Ex \lor Fx)$

II. Where $\mathcal{T}$ is any sentence, an interpretation is often described as a **model** of $\mathcal{T}$ (and $\mathcal{T}$ is thus said to **have a model**) if and only if $\mathcal{T}$ is true on that interpretation. Where $\mathcal{P}_1, \mathcal{P}_2, \ldots, \mathcal{P}_n$ are any sentences, an interpretation is said to be a **model** of $\mathcal{P}_1, \mathcal{P}_2, \ldots, \mathcal{P}_n$ if and only if *all* are true on that interpretation. Using this new terminology, com-

---

[14]P. Bernays and M. Schönfinkel, "*Zum Entscheidungsproblem der mathematischen Logik*", *Mathematische Annalen*, vol. 99 (1928), pp. 342–372.

plete the following:

The sentence $\mathscr{T}$ is a logical truth if and only if each _____ of $\mathscr{T}$ is a _____ of $\mathscr{T}$.

The sentence $\mathscr{T}$ is satisfiable if and only if $\mathscr{T}$ has a _____.

The sentences $\mathscr{T}$ and $\mathscr{U}$ are equivalent if and only if each _____ of _____ is a _____ of _____, and each _____ of _____ is a _____ of _____.

The argument with premises $\mathscr{P}_1, \mathscr{P}_2, \ldots, \mathscr{P}_n$ and conclusion $\mathscr{C}$ is valid if and only if each interpretation of $\mathscr{P}_1, \mathscr{P}_2, \ldots, \mathscr{P}_n$, which is a _____ of _____ is a _____ of _____.

*5.4 Identity and Definite Descriptions

Among dyadic relations, the most important for our purposes is that of **identity**. We shall abbreviate the expression 'x is identical with y' not by 'Ixy' but in the more customary fashion '$x = y$' (we shall also write its negation as '$x \neq y$' instead of '$\sim x = y$'). Identity explicates one of the three important uses of the verb 'to be'. The other two have already been accounted for and are mentioned here briefly. First we have the *existential* sense of 'is', as in

There is a man whom all women admire.

Clearly 'is' here has the force of 'exists' and is handled by means of the existential quantifier. Second, there is the *predicative* (or *copulative*) sense of 'is' — the role it plays, for example, in a singular sentence where a property is ascribed to an individual:

Aristotle is wise.

The verb here plays no essential role as far as logic is concerned — it is an integral part of the predicate schema '... is wise', and no special sign is needed for it in our notation. Hence the above is simply translated 'Wa'.

The third sense is that of identity; 'is' here means 'is identical with'. In ordinary discourse, the 'is' of identity commonly has two singular terms flanking it. Singular sentences of this form will thus be true just in case the very same individual is designated or referred to by both terms. Examples would be:

(1) Eric Blair is George Orwell

(2) Muhammad Ali is Cassius Clay

(3) Taiwan is Formosa

(4) Bonn is the capital of West Germany

(5) The author of *Six Crises* is the 37th President of the U.S.

(6) The greatest hockey player is Gordie Howe

(7) The product of 4 and 16 is the cube root of 262,144

Each of these is true because its component singular terms refer to one and the same individual — which is *not* to say that the sentence is about the component *terms*. (1) is no more *about* the terms 'Eric Blair' and 'George Orwell' than 'Boston is a city' is about the term 'Boston'; rather, the subject matter of (1) is the co-designated man who, for example, wrote the novel *1984*. Moreover, though each sentence in the above list is true, it should not be inferred that its component terms have the same *meaning*. Clearly the terms in (7) convey quite different meanings, though with some minimal arithmetic knowledge it is easily determined that they have the same number as their *reference* and that (7) is hence true. It is in this way that an identity sentence may be true — an individual thus being designated twice over — without also being trivial, as is the case with the uninformative '64 = 64'. Though one and the same individual is designated by 'the product of 4 and 16' and by 'the cube root of 262,144', it is specified or singled out differently in the two cases by virtue of the distinct meanings the terms convey. Or again, while for hundreds of years everyone knew the meanings of 'the Morning Star' and 'the Evening Star', it took considerable research before astronomy had advanced to the point where it was realized that the terms referred to the same individual, and hence that the sentence

The Morning Star is the Evening Star

is true.

The identity relation exhibits three important properties worth noting. First, a dyadic relation — here represented by '*Rxy*' — is said to be **symmetric** if and only if the following holds of it:

$(x)(y)(Rxy \supset Ryx).$

For example, '… is married to ---' expresses a symmetric relation but '… is husband of ---' does not. Identity is obviously symmetric, for given any x and y, if $x = y$ then $y = x$. A relation is said to be **transitive** if and only if:

$(x)(y)(z)[(Rxy \ \& \ Ryz) \supset Rxz].$

The relation represented by '… is larger than ---' is transitive, but that by '… is mother of ---' is not. Identity is also transitive, since if $x = y$ and $y = z$ then $x = z$. Finally, a relation is said to be **reflexive** if and only if

$(x)(Rxx),$

that is, every individual bears R to itself. Though identity is reflexive (every substitution instance of '$x = x$' is true and necessarily so), the relation expressed by '...is taller than---' clearly is not.[15]

Sentences (1), (2) and (3) all contain proper names flanking 'is', and using obvious notation are expressed:

(1') $e = g$

(2') $m = c$

(3') $t = f.$

Sentences (4) through (7), however, each contain at least one definite description. We have yet to discuss the translation of such terms, and shall temporarily put off further discussion of them.

Identity enables us to express many arithmetical sentences that heretofore could not be handled, for example

(8)   There are at least two spies in the room.

Now, we know that 'There is at least one spy in the room' is expressed simply as '$(\exists x)(Sx)$', so it might be thought that (10) is adequately rendered as

$(\exists x)(\exists y)(Sx \mathbin{\&} Sy).$

---

[15]Among relations which are *not* symmetric some are **asymmetric**, others not. A relation is asymmetric if and only if:

$(x)(y)(Rxy \supset {\sim}Ryx).$

Thus '...is husband of---' expresses an asymmetric relation, but '...loves---' expresses one which is neither symmetric nor asymmetric. Similarly, there are relations that are merely *not* transitive and those that are **intransitive**:

$(x)(y)(z)[(Rxy \mathbin{\&} Ryz) \supset {\sim}Rxz]$

as may be seen by contrasting '...loves---' with '...is mother of---'. Likewise, we may contrast relations that are merely not reflexive (again '...loves---') with those which are **irreflexive** (for example, '...is more important than---'):

$(x)({\sim}Rxx).$

However, this will not do, for suppose that in fact only one spy, whom we shall designate as '*a*', is in the room; then (8) is *false* but the above is *true* since the substitution instance '*Sa & Sa*' is true. This sentence of our notation, then, does not provide the desired translation, since it allows that *x* and *y* might be the same individual. To remove this possibility, it is necessary to add the further conjunct '$x \neq y$', so that (8) becomes

(8') $(\exists x)(\exists y)(Sx \mathbin{\&} Sy \mathbin{\&} x \neq y).$[16]

By the same token, the sentence

(9)   There are at least three spies in the room,

or in other words 'There are three or more spies in the room', may be expressed in symbols as

(9') $(\exists x)(\exists y)(\exists z)(Sx \mathbin{\&} Sy \mathbin{\&} Sz \mathbin{\&} x \neq y \mathbin{\&} y \neq z \mathbin{\&} x \neq z).$

Sentences asserting the existence of at least four, five, and more spies may be handled in similar fashion.
Now consider the sentence

(10)   There is a senator who can outtalk everyone.

A moment's reflection will show that

$(\exists x)[Sx \mathbin{\&} (y)(Py \supset Oxy)]$

does not do (10) full justice as it is normally understood, for if *every* person is outtalked by the senator then he outtalks himself as well, which is hardly plausible even for senators. The force of 'everyone' in (10) is that of 'everyone *else*', and hence the appropriate translation would be

(10') $(\exists x)\{Sx \mathbin{\&} (y)[(Py \mathbin{\&} y \neq x) \supset Oxy]\}.$

We have already considered many types of sentence containing the

---

[16]For brevity, we sometimes drop interior parentheses when no ambiguity results.

troublesome word 'only'; yet a further example is

(11)   Al is the only spy in the room.

(11) is best understood as claiming, first, that Al is a spy in the room, and second, nothing else is. Thus we have the conjunction

(11') $Sa \ \& \ (x)(x \neq a \supset \ \sim Sx)$.

Sentences containing the expressions 'at most' and 'no more than' may also be expressed using the identity sign. Consider:

(12)   No more than one spy is in the room

(13)   At most two spies are in the room.

Both sentences would generally be understood as *not* claiming the existence of at least one spy despite the fact that they may be reformulated to read, 'There is (are) at most...'. In (12), the claim amounts to: *if* there is a spy in the room, he's the only one. So (12) becomes

(12') $(x)(y)[(Sx \ \& \ Sy) \supset x = y]$.

Similarly, (13) is expressed

(13') $(x)(y)(z)[(Sx \ \& \ Sy \ \& \ Sz) \supset (x = y \lor y = z \lor x = z)]$.

In other words, if x, y, and z are all spies, then at least "two" of them are the same.

Now that we have obtained translations of (8) and (13), we are able to handle:

(14)   There are two spies in the room
       Exactly two spies are in the room.

To say there are *exactly* two is to say there are at least two *and* at most two. So (14) may be translated using the tools provided by (8) and (13):

(14') $(\exists x)(\exists y)\{Sx \ \& \ Sy \ \& \ x \neq y \ \& \ (z)[Sz \supset (z = x \lor z = y)]\}$.

The last conjunct in effect ensures that the distinct spies x and y are the *only* ones. A less complicated example would be

(15)   There is exactly one deity,

which accordingly becomes

(15') $(\exists x)[Dx \,\&\, (y)(Dy \supset y = x)]$.

Definite descriptions were briefly introduced in Section 5.1 as a variety of singular term in addition to that of proper names. Instead of naming the designated individual, such a term uniquely describes it. Thus in (4) above, 'the capital of West Germany' is a definite description, and the use of 'the' in the singular is characteristic of such expressions (though they are sometimes abbreviated in idiomatic speech to the point where 'the' disappears, for example 'Al's father'). When 'the' is not used in the singular, the descriptive phrase is not definite — it does not purport to refer descriptively to just *one* individual. Thus, 'the men in the next room' is not a definite description, but in 'the capital of West Germany' we have an expression designating the one and only one individual truly described as 'capital of West Germany'.[17]

Since it has a descriptive dimension, any definite description must have at least one general term and hence may be analyzed in terms of an open sentence. Thus, for 'the capital of West Germany', we have 'x is capital of West Germany', and the use of 'the' carries with it the presumption of exactly one individual so described. We may thus reformulate the description as

the (individual) x such that x is capital of West Germany

and we shall abbreviate the phrase 'the x such that ...' by attaching an inverted iota to 'x' and enclosing both in parentheses, thus giving '$(\imath x)$'. We shall call this sign the **iota operator**, and attaching it to an open sentence thus provides us with a means of translating definite descriptions

---

[17]It should be noted that there are expressions that grammatically resemble definite descriptions but are best construed as proper names, such as 'the Windy City', 'the Sunshine State', 'the Empire State Building'. Here, the use of capital letters indicates that the designated individuals are not being uniquely described as such, but are named — albeit through the use of terms felt to be especially characteristic of their bearers ('windy' of Chicago, 'sunshine' of Florida). In fact, such expressions often serve as nicknames.

into our notation. So using 'Cx' for the above open sentence, the description is rendered

$(\imath x)(Cx)$.

The iota operator is *not* a quantifier — when attached to an open sentence the result is not a sentence true or false, but a descriptive singular *term*. We may now return to (4):

(4') $b = (\imath x)(Cx)$.

Using 'Wx' for 'x wrote *Six Crises*' and 'Px' for 'x is 37th President of the U.S.', (5) becomes:

(5') $(\imath x)(Wx) = (\imath y)(Py)$

Now (6) contains the superlative term 'greatest' and such terms may be handled as follows: in speaking of a greatest hockey player, x, we mean that x is a hockey player greater than any *other* hockey player, i.e. '$Hx \,\&\, (y)[(Hy \,\&\, y \neq x) \supset Gxy]$'. So we may express (6) as

(6') $(\imath x)\{Hx \,\&\, (y)[(Hy \,\&\, y \neq x) \supset Gxy]\} = g$.

Importing now some arithmetic vocabulary, (7) may be expressed

(7') $(\imath x)(x = 4 \times 16) = (\imath y)(y = \sqrt[3]{262{,}144})$.

A description, like a constant, may replace a free variable in an open sentence to yield a true-or-false sentence as substitution instance. Thus, where 'a' is 'Aristotle' and '$(\imath x)(Axd)$' is 'the author of *De Anima*', both terms may replace the 'x' of 'Px' ('x is a philosopher') thus giving

Aristotle is a philosopher Pa
The author of *De Anima* is a philosopher $P(\imath x)(Axd)$

Thus, singular descriptions of the form '$(\imath x)(\ldots x \ldots)$' would seem to behave no differently from the individual constants of our symbolic notation. Yet there is an important disparity that requires us to push further our analysis of sentences containing descriptions. Our use of individual constants thus far has proceeded on the supposition that in any context where it occurs, a constant always designates a particular individual of the presumed universe of discourse. However, this supposition cannot be extended to the singular descriptions of our notation, for it should be apparent that we can form at will such descriptions that fail to designate or refer to any individual at all. For example:

($\imath x$)(x hit 65 home runs in one major league season)
($\imath x$)(x hit 55 home runs in one major league season)

The first fails to refer because *no* individual has hit 65 home runs, the second because *more than one* has hit 55 home runs. We can easily multiply such examples:

($\imath x$)(x is a man who has run a two-minute mile)
($\imath x$)(x is monarch of France)
($\imath x$)(x has squared the circle)
($\imath x$)(x is a senator from Indiana)

We are thus left with a problem concerning the analysis of sentences with singular descriptions, for we must now face the possibility of reference failure. Keeping this in mind, how are we to analyze such sentences? In particular, under what conditions will they be true and under what false? Let us consider a sentence with a nonreferring description:

(16)   The man who has run a two-minute mile is bald,

or in other words, where 'Tx' is 'x is a man who has run a two-minute mile' and 'Bx' is 'x is bald':

(16′) $B(\imath x)(Tx)$.

One answer to the above questions is to maintain that such a sentence has no truth-value, since the presupposition of unique reference is not satisfied. This accords well with our intuitions, for if someone were (seriously) to assert (16) our response would very likely take the form of denying that there is such a man, *not* of claiming that (16) is false — which would carry the suggestion that there *is* such a man but that he is not bald. To say that (16) has no truth-value is not, according to this view, to say that it has no meaning. Rather, it makes perfectly good sense in that it is easily understood by any fluent speaker of English. The claim is that for (16) — or (16′) — to be considered a candidate for *either* truth *or* falsity a certain presupposition must be satisfied:

(17)   There is exactly one man who has run a two-minute mile.

(17′) $(\exists x)[Tx \;\&\; (y)(Ty \supset y = x)]$.

Since, however, (17) is false, the question of the truth-value of (16) does not arise. Only if (17) is itself true will (16) be considered true or false — true if the designated individual *is* bald, false if he is not.[18]

However tempting this approach may be, we shall not pursue it any further here. Natural though it is to regard sentences with nonreferring descriptions as lacking a truth-value, the practice has serious drawbacks with respect to the evaluation of arguments containing such sentences in our notation. Our accounts of validity, equivalence, and the like clearly hinge on sentences having a truth-value on *all* interpretations, and were we now to remove this condition, new and highly complex definitions would be required — definitions the adequacy of which is still a matter of intense philosophic controversy. We will do better, then, if we can find an account of sentences with definite descriptions that, first, ascribes to them a truth-value on all interpretations, and second, accords reasonably well with our intuitions concerning their use in ordinary discourse and their behavior in deductive inference.

Such an approach has been offered by Bertrand Russell and has long had wide acceptance.[19] On this view, (17) is not a presupposition for (16) having a truth-*value*, but simply for its being true — (17) in effect gives us part of the truth-conditions for (16). On Russell's view, then, (16) *asserts* the existence of exactly one man who has run a two-minute mile instead of presupposing it as a condition for having a truth-value. More fully,

---

[18]For more on this approach, see Gottlob Frege, "On Sense and Reference," in P. Geach and M. Black (Eds.), *Philosophical Writings of Gottlob Frege*, Oxford: Blackwell, 1952. In his formal language, however, Frege simply assigns a reference to what would otherwise be nonreferring descriptions. A different though related approach may be found in P. F. Strawson, "On Referring," in A. Flew (Ed.), *Essays in Conceptual Analysis*, London: Macmillan, 1956. Both essays are classics and have been reprinted in many other anthologies on logic and language.

[19]Bertrand Russell, "On Denoting," *Mind*, vol. 14 (1905), pp. 479–493. Reprinted in many places, among them: H. Feigl and W. Sellars (Eds.), *Readings in Philosophical Analysis*, New York: Appleton-Century-Crofts, 1949; and I. Copi and J. Gould (Eds.), *Contemporary Readings in Logical Theory*, New York: Macmillan, 1967 (the latter contains the Frege and Strawson essays as well). See also Russell's *Introduction to Mathematical Philosophy*, London: Allen and Unwin, 1920, Chapter 16, and his *Principia Mathematica* (with A. N. Whitehead), vol. I, Cambridge: Cambridge Univ. Press, 1910.

Russell claims (16) will be true if and only if:

  i. at least one man has run a two-minute mile,
 ii. at most one man has run a two-minute mile, and
iii. that man is bald.

Hence, according to Russell, (16) is *false*, since it has the same truth-conditions as the existentially quantified sentence

(16″) $(\exists x)[Tx \ \& \ (y)(Ty \supset y = x) \ \& \ Bx]$.

More generally, a sentence containing a definite description and therefore one that has been provisionally expressed in our notation as being of the form

$--- (\imath x)(\ldots x \ldots) ---$

will hereafter be regarded as equivalent to one of the form

$(\exists x)[\ldots x \ldots \ \& \ (y)(\ldots y \ldots \supset y = x) \ \& \ ---x---]$.

On this analysis, a sentence purporting to uniquely describe an individual and ascribe a property to it will be false when either:

(a) the description fails to refer (either because there is no individual so described or because there are two or more), or
(b) the description does refer but the individual lacks the property in question.

Thus (16) would also be false, according to (b), if there were a man who has run a two-minute mile but who is not bald. Following Russell, then, we shall translate (16) *and* (16′) as (16″), thus dispensing with the iota operator and expressions of the form '$(\imath x)(\ldots x \ldots)$' for the balance of our discussion. Notice that upon adopting this course, singular descriptions will no longer occur in our notation. Sentences like (16) and (16′) are now reexpressed in such a way that there is no *one* expression or phrase in their translation—as in (16″)—that replaces the description; instead, the entire sentence does the job. Russell describes this as the *contextual elimination* of definite descriptions; they are not defined away in the sense in which, say, 'bachelor' is by 'unmarried man', but rather defined within the sentential context as a whole. In going from (16) or (16') to (16"), the description 'the man who has run a two-minute mile' ('$(\imath x)(Tx)$') simply disappears, its role taken over by the entirety of (16").

Let us now consider the ostensible negation of (16) and of (16'):

(18)   The man who has run a two-minute mile is not bald.

(18') $\sim B(\imath x)(Tx)$.

Since (16) and (16') can each be *false* in two different ways, clearly (18) and (18') can be *true* in two different ways—either if (a) holds or if (b) holds. Since (a) and (b) are distinct conditions for truth, it follows that (18) and (18') on Russell's theory can both be understood in two different ways. Understood one way, they will be true if and only if there *is* just one man who has run a two-minute mile but who is *not* bald, and on this reading the descriptions 'the man who has run a two-minute mile' and '$(\imath x)(Tx)$' are said to have **primary occurrence** in (18) and (18') respectively. Understood the other way, they will be true if no man or more than one man has run a two-minute mile, and on this second reading the descriptions are said to have **secondary occurrence** in (18) and (18'). The expressions (18) and (18') must therefore, according to Russell, be regarded as ambiguous. If the component descriptions are taken to have primary occurrence, (18) and (18') are both symbolized

(18p) $(\exists x)[Tx \ \& \ (y)(Ty \supset y = x) \ \& \sim Bx]$.

When the descriptions in (18) and (18') are taken as having primary occurrence, then a condition for the sentence's being true is that there *be* an individual referred to. When the descriptions are regarded as having secondary occurrence, however, the sentences may be true even if there is no individual designated by those descriptions. Hence, a quite different translation is required when the descriptions have secondary occurrence. Here, (18) and (18') must be understood to mean

(18s) $\sim(\exists x)[Tx \ \& \ (y)(Ty \supset y = x) \ \& \ Bx]$.

We may now translate (4), (5), and (6) in accordance with Russell's technique. (4) says that there is exactly one x which is capital of West Germany and is (identical with) Bonn. Thus we have:

(4") $(\exists x)[Cx \ \& \ (y)(Cy \supset y = x) \ \& \ x = b]$

(5") $(\exists x)(\exists y)[Wx \ \& \ (z)(Wz \supset z = x) \ \& \ Py \ \& \ (w)(Pw \supset w = y) \ \& \ x = y]$

In (5″), the first two conjuncts say in effect that exactly one individual wrote *Six Crises* and the second pair say that exactly one is president of the U.S. The fifth conjunct ensures their identity. (6) now becomes

(6″) $(\exists x)\{Hx \,\&\, (y)[(Hy \,\&\, y \neq x) \supset Gxy] \,\&\, x = g\}.$

(7) is left as an exercise for the patient reader.

EXERCISES

I. Put each of the following in symbols using the suggested notation. Use Russell's method on those containing definite descriptions. The identity sign is required for all.

1. There are two gods at most. (Gx-x is a god.)
2. There is at least one but no more than three gods. (Same notation as (1).)
3. The man who defeated Dempsey fought only heavyweights. (Mx-x is a man; Dxy-x defeated y; d-Dempsey; Fxy-x fought y; Hx-x is a heavyweight.)
4. Nelson's most recent speech was untelevised. (n-Nelson; Sxy-x is a speech given by y; Rxy-x is more recent than y; Tx-x was televised.)
5. The only man who hated every child is W. C. Fields. (Mx-x is a man; Hxy-x hates y; Cx-x is a child; f-W. C. Fields.)
6. Cal and Dick were the only suspects who confessed. (c-Cal; d-Dick; Sx-x is a suspect; Cx-x confessed.)
7. There are exactly two reasons for watching Raquel. (Rxy-x is a reason for watching y; r-Raquel.)
8. Only the best applicant will be hired. (Ax-x is an applicant; Bxy-x is better than y; Hx-x will be hired.)
9. Gerry, Dave, and at least two other suspects were indicted. (g-Gerry; d-Dave; Sx-x is a suspect; Ix-x was indicted.)
10. Only Audrey hit Frank on his arm. (a-Audrey; f-Frank; Axy-x is an arm of y's; $Hxyz$-x hit y on z.)
11. Audrey hit only Frank on his arm. (Same notation as (10).)
12. Audrey hit Frank on his only arm. (Same notation as (10).)
*13. Mary and Jane are twins. (m-Mary; j-Jane; Tx-x is a time; $Bxyz$-x gave birth to y at z.)

*II. Express
    I want a girl just like the girl that married dear old Dad
such that, first, the description 'the girl that married dear old Dad' has primary occurrence, and second, has secondary occurrence. (Gx-x is a girl; Wx-I want x; Lxy-x is just like y; Mxy-x married y; d-dear old Dad.)

CHAPTER SIX

QUANTIFICATIONAL LOGIC
Deductive Methods

6.1 The Rules EI, EG, UI, UG, and QN

Certain arguments whose premises and conclusions involve quantifiers can be shown valid with no more deductive apparatus than is available from Chapter 3:

1. $(\exists x)(Px) \supset (Gae \supset Laf)$	Premise	
2. $\sim Laf$	Premise	DEDUCE: $Gae \supset \sim(\exists x)(Px)$
3. $(\exists x)(Px)$	Assumption	
4. $Gae \supset Laf$	1, 3 MP	
5. $\sim Gae$	2, 4 MT	
6. $(\exists x)(Px) \supset \sim Gae$	3–5 CP	
7. $\sim\sim Gae \supset \sim(\exists x)(Px)$	6 Trans	
8. $Gae \supset \sim(\exists x)(Px)$	7 DN	

But if we restrict ourselves to just those earlier rules we will not be able to construct a deduction even for the obviously valid argument:

1a. $(\exists x)[Px \& (Gxe \& Lxf)]$	There exists at least one prime number greater than 8,000,000 and less than $8,000,000! + 2$
$\therefore (\exists x)(Px \& Gxe)$	Therefore there exists at least one prime number greater than 8,000,000.

The scope of the quantifier in (1a), unlike that of the quantifier in (1), extends across the whole sentence. For that reason (1a) is clearly not a conjunction, and so neither **Simp** nor any of the other rules of Chapter 3 apply.

Of course if we were entitled to suppose that 8,000,001 ('*a*' for ease of reference) is a prime number greater than 8,000,000 and less than 8,000,000! + 2 — that is if we were entitled to

2a. *Pa & (Gae & Laf)*,

we could go on to argue

3a. *(Pa & Gae) & Laf*   2a Assoc
4a. *Pa & Gae*   3a Simp
5. *(∃x)(Px & Gxe)*

intending to justify the move from (4a) to (5) by noting that any argument of the form

...*a*...
∴ *(∃x)(...x...)*

must be valid (if *a* has a certain property, then obviously *something* does). Or if we could infer that 8,000,002 ('*b*' here) is a prime number between 8,000,000 and 8,000,000! + 2, that is, if we had

2b. *Pb & (Gbe & Lbf)*,

we could argue

3b. *(Pb & Gbe) & Lbf*   2b Assoc
4b. *Pb & Gbe*   3b Simp
5. *(∃x)(Px & Gxe)*

Indeed, if we knew of any particular individual *c* that is a prime number between 8,000,000 and 8,000,000! + 2, we could argue from

2c. *Pc & (Gce & Lcf)*

to our conclusion:

3c. $(Pc \ \& \ Gce) \ \& \ Lcf$ 2c Assoc

4c. $Pc \ \& \ Gce$ 3c Simp

5. $(\exists x)(Px \ \& \ Gxe)$

But none of this, of course, provides us with a *deduction* of (5) from (1a), for neither (2a) nor (2b) nor anything like (2c) follows from (1a). Our premise does not tell us *which* thing or things (there may be several) are prime numbers between 8,000,000 and 8,000,000! + 2 and so does not entitle us to any of these steps. The difficulty seems to be that (1a) tells us that there exists at least one number of the sort in question but gives us no name for it, and so leaves us with no way to refer to it or talk about it.

In ordinary intuitive reasoning we often meet this sort of situation in a direct and straightforward way: given no name for a thing we are assured exists and need to refer to, we provide a temporary one by means of a locution like "let v be one such" or "pick one, and call him 'John Doe'." To handle the argument considered here, for example, it would be natural to argue along the following lines:

1. Our premise assures us that there exists at least one prime number greater than 8,000,000 and less than 8,000,000! + 2.

2. Let v be one such; that is, pick any prime number greater than 8,000,000 and less than 8,000,000! + 2 and call it 'v' (if there are several such numbers, it won't matter which one is chosen).

3. Then of course v is a prime number greater than 8,000,000 as well as being less than 8,000,000! + 2.

4. So v is (among other things) a prime number greater than 8,000,000.

5. Consequently there does exist at least one prime number greater than 8,000,000 (v, for example, whichever number it happens to be).

To make use of this device in our deductive system, we will agree to keep the hitherto unused letters 'u', 'v', 'u''', 'v''' and so forth (which we will call **pseudo-names**) available for use only in situations like this one, where we want to refer to a thing of a certain sort but do not know or care *which* thing of that sort it is. (In the example above we are told that there are prime numbers between certain limits, but are not told which numbers between those limits are prime.) Thus our pseudo-names 'u', 'v', and so forth, contrast sharply with individual constants such as 'a', 'b', and 'c', which are used to designate specific individuals just as genuine proper names ('Aristotle', for example) are.

Then our deduction can be obtained directly from the informal argument (1)–(5) above:

1. $(\exists x)[Px \,\&\, (Gxe \,\&\, Lxf)]$ Premise   DEDUCE: $(\exists x)(Px \,\&\, Gxe)$

2. $Pv \,\&\, (Gve \,\&\, Lvf)$

3. $(Pv \,\&\, Gve) \,\&\, Lvf$ 2 Assoc

4. $Pv \,\&\, Gve$ 3 Simp

5. $(\exists x)(Px \,\&\, Gxe)$

The rule permitting us to move from *existentially* quantified open sentences like (1) to such *instances*[1] of them as (2) will be called **Existential Instantiation** (**EI**, for short) and it can be formulated most generally as follows:

Existential Instantiation (EI)

Where β is a pseudo-name (*not* a constant), one may in a deduction infer from a previously obtained existentially quantified sentence

$(\exists\mu)(\ldots\mu\ldots)$

any corresponding sentence

$\ldots\beta\ldots$

provided that: i. β does not occur in any *earlier* line of the deduction;
ii. $\ldots\beta\ldots$ results from $\ldots\mu\ldots$ by replacing each occurrence of μ with β (making *no* other changes).[2]

---

[1]We call expressions which involve occurrences of pseudo-names (for example, '*Pv*' and '*Gve & Lvf*') *sentences* because of their similarity to such English sentences as 'It hit me on the head'. Often the context in which the sentence 'It hit me on the head' is used is all that makes clear the antecedent reference of the pronoun 'it'. Imagine that we rush into a room shortly after a number of cans have fallen from a shelf and find a friend surrounded with cans and cluthching his head. "Did one of them hit you?" we might ask, and he could reply truly "Yes, *it* hit me on the head" even though none of us may have any way of finding out exactly *which* can it was that hit him there.

To footnote 12 of the preceding chapter, we now add:

v. The result of replacing one or more constants in a sentence with a pseudo-name ('*u*', '*u′*', '*v*', '*v′*', etc.) is a sentence.

[2]We shall follow the practice throughout of employing only members of the *v*-family ('*v*', '*v′*', '*v″*', and so on) in applications of **EI**. Though not required, this practice will pay off later when we come to the rule called **UG** for which we are reserving pseudo-names in the *u*-family.

The reason for the restriction to pseudo-names should be clear from our preceding discussion. With proviso (i) we forestall unjustifiable attempts to treat several presumably different things as if they were one and the same by giving the same name to all of them, as in:

1. $(\exists x)(Cx)$	Premise	There are some cats
2. $(\exists x)(Dx)$	Premise	There are some dogs
3. Cv	1 EI	Let v be a cat
4. Dv	2 EI (mistake)	Let v be a dog (too?)
5. Cv & Dv	3, 4 Conj	Then v is *both* a cat *and* a dog
6. $(\exists x)(Cx$ & $Dx)$		So there is something that is both a cat and a dog.

The premises are both true but the conclusion reached is obviously false, so there must be a mistake somewhere and it's not hard to find. If we're sure there are cats (as premise one tells us) we may call one of them 'v'. Or if we're also sure there are dogs (as premise two tells us) we may prefer to call one of *them* 'v' instead. But we cannot justifiably use 'v' as a name for a cat at one point and then later, in the same context, decide to use 'v' as a name for a dog, too — doing so would be tantamount to supposing that there was some one thing that was both, and our premises give us no right to *that* supposition. The best we can legitimately do at line (4) is get 'Dv''' or 'Dv'''', using a *different* name for the dog than we used for the cat; and of course neither of these will lead to (6).

We need proviso (ii) as well, to rule out attempts to infer things like 'Pv & Gxe' and 'Pv & Gvv' from '$(\exists x)(Px$ & $Gxe)$'.

EI, then, gives us a way to get existential quantifiers *out* of deductions. But after we've gotten them out we often later want to put them back *in*, as we did earlier when we wanted to move from

Pv & Gve

to

$(\exists x)(Px$ & $Gxe)$.

The rule needed here is:

Existential Generalization (EG)

Where α is a pseudo-name *or* constant, one may in a deduction infer from a previously obtained sentence

...α...

any corresponding existentially quantified sentence

$(\exists\mu)(\ldots\mu\ldots)$

provided that: i. μ is a variable not already occurring in $\ldots\alpha\ldots$;
               ii. $\ldots\mu\ldots$ results from $\ldots\alpha\ldots$ by replacing *at least one* occurrence of α with μ (making *no* other changes).

The inference from

1. *Fba* v $(\exists x)(Fxb)$ Either Brown will fire Adams or someone[3] will fire Brown

to

2. $(\exists y)[Fya$ v $(\exists x)(Fxy)]$ Someone is such that either he will fire Adams or someone will fire him

is clearly valid so we permit **EG** on individual constants as well as on pseudo-names. Of course the move from (1) to

3. $(\exists y)[Fya$ v $(\exists x)(Fxb)$ Someone is such that either he will fire Adams or someone will fire Brown

is also valid, and that's why proviso (ii) permits replacement of only *some* of the occurrences of the pseudo-name or constant involved. *No* other changes are permitted, naturally — we can no more justifiably move from (1) to

4. $(\exists y)[Fyy$ v $(\exists x)(Fxy)]$ Someone is such that either he will fire himself or someone will fire him

then we could move to '$(\exists y)[Qya$ v $(\exists x)(Zxy)]$'. And of course the restriction in (i) is necessary to prevent such undesirable moves as that from (1) to

5. $(\exists x)[Fxa$ v $(\exists x)(Fxx)]$,

for the latter would be totally ambiguous. (Does it say that there's some-

---

[3]For this example, we restrict our universe of discourse to persons.

one who will either fire Adams or be himself fired by someone? Or that there's someone who will either fire Adams or fire someone? Or that there's someone who will either fire Adams or fire himself? Or that there's someone who's such that either he will fire Adams or else someone will fire himself?) To apply **EG** to (1) one has to pick a "new" variable not already occurring in it, as was done, for example, by choosing 'y' in passing from (1) to (2).

Undoubtedly

6. $(\exists y)(Fya) \lor (\exists x)(Fxb)$ Either someone will fire Adams or someone will fire Brown

also follows from (1) but it ought to be noticed at the outset that **EG** does *not* permit us to go from (1) to (6) *directly* — **EG** only permits the existential quantifier to be inserted in front of a whole line of a deduction and does not allow us to insert quantifiers into the middle. To get (6) from (1) we would need something like the following deduction:

1. $Fba \lor (\exists x)(Fxb)$	Premise	DEDUCE: $(\exists y)(Fya) \lor (\exists x)(Fxb)$
2. $\sim(\exists x)(Fxb)$	Assumption	
3. $(\exists x)(Fxb) \lor Fba$	1 Com	
4. Fba	2, 3 DS	
5. $(\exists y)(Fya)$	4 EG	
6. $\sim(\exists x)(Fxb) \supset (\exists y)(Fya)$	2–5 CP	
7. $\sim\sim(\exists x)(Fxb) \lor (\exists y)(Fya)$	6 Impl	
8. $(\exists y)(Fya) \lor \sim\sim(\exists x)(Fxb)$	7 Com[4]	
9. $(\exists y)(Fya) \lor (\exists x)(Fxb)$	8 DN	

EXERCISES

Using the rules **EI** and **EG** in addition to those of Chapter 3, deduce the conclusions of the following arguments from their premises:

    1. $(\exists x)(Fx \& Gx)$
          $\therefore (\exists x)(Gx)$

---

[4]In accordance with our agreement in Section 3.4, this application of **Com** can be combined with the ensuing application of **DN** into a single step. In our examples in this section and the next, we will continue listing each application of each rule on a separate line for the sake of clarity; after we've acquired some experience with the quantification rules, however, we will drop the separate listings even in our examples.

2. $(\exists x)(Fx)$
   $\therefore (\exists y)(Fy)$

3. $(\exists x)(Gx \ \& \ Hx)$
   $\therefore (\exists z)(Hz)$

4. $(\exists x)(Fx \ \& \ Gx)$
   $(\exists x)(Fx \ \& \ Hx)$
   $(\exists y)(Gy \ \& \ Fy) \supset (\exists x)(Gx \ \& \ Hx)$
   $\therefore (\exists y)(Gy \ \& \ Hy)$

5. $(\exists x)(Lxx)$
   $\therefore (\exists x)(\exists y)(Lxy)$

6. $(\exists x)(Bx) \supset Be$
   $\therefore (\exists x)(Bx) \equiv Be$

7. $(\exists x)(Rxx) \supset Rcc$
   $\therefore Rcc \equiv (\exists y)(Ryy)$

8. $(x)(Fx \supset Gx) \supset [\sim(\exists y)(Gy) \supset \sim(\exists z)(Fz)]$
   $(\exists z)(Fz)$
   $\therefore (x)(Fx \supset Gx) \supset (\exists y)(Gy \lor Hy)$

9. $\sim(\exists x)(Px \ \& \ \sim Wx)$
   Pa
   $\therefore Wa$

*10. $\sim(\exists x)(\sim Ix)$
   $\sim(\exists x)(\sim Wx)$
   $\therefore \sim(\exists x)(\sim Ix \lor \sim Wx)$

EG and **EI** enable us to get existential quantifiers into and out of deductions, and it will be extremely convenient to have a similar pair of rules for eliminating and restoring universal quantifiers as well. The desirability of having a rule of **Universal Instantiation (UI)** permitting us to move from *universally* quantified sentences to particular substitution instances of the open sentences within them, for example, is evident when we consider the argument:

1. $(x)(Px \supset Wx)$	Premise	All philosophers are wise
2. Pa	Premise	Aristotle is a philosopher
$\therefore Wa$		Therefore Aristotle is wise.

The first premise tells us that *all* things that are philosophers are things that are wise and so tells us something about each thing in the universe. It clearly follows from (1), for example, that if Descartes is a philosopher then Descartes is wise, that if the Columbia River is a philosopher then the Columbia River is wise, that if Bozo the Clown is a philosopher then Bozo the Clown is wise, and so on through a variety of other such conditionals including one of special interest to use here:

3. *Pa ⊃ Wa*   If Aristotle is a philosopher then Aristotle is wise.

Once we have (3), of course, we can complete our deduction:

4. *Wa*   2, 3 MP

What is true of every individual naturally is true of any particular individual we happen to be interested in, so the move from (1) to (3) is clearly valid, as are each of the following arguments:

$(x)(Ix \& Wx)$	$(x)(Lxx)$	$(x)(\exists y)(Ax \supset By)$	$(x)(y)(Rxy)$	$(y)(Rhy)$
∴ *Ib & Wb*	∴ *Ldd*	∴ $(\exists y)(Ae \supset By)$	∴ $(y)(Rhy)$	∴ *Rhh*

So we want to formulate a rule like this:

Universal Instantiation (UI)
Where α is a pseudo-name *or* constant, one may in a deduction infer from a previously obtained universally quantified sentence

$(\mu)(\ldots \mu \ldots)$

any corresponding sentence

$\ldots \alpha \ldots$

provided that $\ldots \alpha \ldots$ results from $\ldots \mu \ldots$ by replacing each occurrence of μ
         with α (making *no* other changes).

Here, we allow instantiation to pseudo-names as well as to constants so that an argument like

$(x)(Px \supset Wx)$	All philosophers are wise
$(\exists x)(Px)$	There exists at least one philosopher
∴ $(\exists x)(Px \& Wx)$	Therefore there exists at least one philosopher who is wise

can be handled along the following natural lines:

1. $(x)(Px \supset Wx)$ Premise

2. $(\exists x)(Px)$ Premise   DEDUCE: $(\exists x)(Px \& Wx)$

3. *Pv*   2 EI

4. *Pv* ⊃ *Wv*   1 UI

5. *Wv*   3, 4 MP

6. *Pv* & *Wv*   3, 5 Conj

7. (∃*x*)(*Px* & *Wx*)   6 EG

(2) tells us that there are some philosophers and we choose, with (3), to use '*v*' to refer to one of them. Now along with telling us things about Aristotle, Bozo the Clown, and the Columbia River, (1) also tells us something about *v*, whatever it may be, so the move from (1) to (4) is clearly acceptable.

It is extremely important to remember that the quantification rules introduced here are rules of *inference*, not rules of replacement, and so can be applied only to quantifiers whose scope extends across the entire line. For example, the argument

(*x*)[*Bx* ⊃ (∃*y*)(~*Gy*)]   Everything is such that if it's blue then something is not green

∴ *Ba* ⊃ (∃*y*)(~*Gy*)   Therefore if the Atlantic Ocean is blue then something is not green

is valid, and **UI** permits passing from its premise to its conclusion. However, the argument

(*x*)(*Bx*) ⊃ ~(∃*y*)(*Gy*)   If everything is blue then nothing is green

∴ *Ba* ⊃ ~(∃*y*)(*Gy*)   Therefore if the Atlantic Ocean is blue then nothing is green.

is clearly invalid, and **UI** does *not* license such inferences.

Our last rule, **Universal Generalization (UG)**, will be designed to permit us, under certain very special conditions, to introduce universal quantifiers into deductions. Sometimes one can establish a sentence of the form

…*β*…

(*β* not a constant but a pseudo-name) in such a way as to make it clear that *all* similar sentences,

…*a*…
…*b*…
…*c*…

and so on, could have been established as well. **UG** will permit us, in such special cases, to infer the general conclusion we would clearly be entitled to:

$(\mu)(\ldots \mu \ldots).$[5]

Consider, for example, the argument:

1. $(x)(Ix)$	Premise	Everything is interesting
2. $(x)(Wx)$	Premise	Everything is worthy of study
$\therefore (x)(Ix \ \& \ Wx)$		Therefore everything is both interesting and worthy of study.

With the rules we already have we could begin in any of several ways, including:

1. $(x)(Ix)$	Premise		1. $(x)(Ix)$	Premise		1. $(x)(Ix)$	Premise
2. $(x)(Wx)$	Premise		2. $(x)(Wx)$	Premise		2. $(x)(Wx)$	Premise
3. Ia	1 UI		3. Ib	1 UI		3. Ic	1 UI
4. Wa	2 UI		4. Wb	2 UI		4. Wc	2 UI
5. $Ia \ \& \ Wa$	3, 4 Conj		5. $Ib \ \& \ Wb$	3, 4 Conj		5. $Ic \ \& \ Wc$	3, 4 Conj

But it is clear that none of these fine beginnings is going to justify our concluding '$(x)(Ix \ \& \ Wx)$'. It does not follow from '$Ia \ \& \ Wa$' that everything

---

[5]The reader may find it helpful to think of a geometer who has established that

If *ABC* is an equiangular triangle
then *ABC* is an equilateral triangle

but wants to conclude not just that the particular triangle shown in the accompanying diagram is equilateral if equiangular but that

All equiangular triangles are equilateral.

Of course there are many quite general methods he might have used to establish the former that would justify his concluding the latter, but obviously a lot depends on the generality of the manner in which he did establish the former. If he did it by first measuring the angles of the particular triangle shown in the diagram, for example, and then measuring its sides, he is clearly *not* justified in leaping to a conclusion concerning *all* triangles.

is interesting and worthy of study, nor does it follow from '*Ib* & *Wb*' nor from '*Ic* & *Wc*'. To be justified in concluding that *everything* is both interesting and worthy of study we must show more than that a number of particular individuals are. What we want to show is that no matter which individual we consider, it is both interesting and worthy of study. We want to say, in effect:

Pick anything you wish, anything at all. Then it follows from (1) that it's interesting and from (2) that it's worthy of study. So of course it – that thing you picked, whatever it was – is both interesting and worthy of study. And since you were free to choose *any* individual whatever, we are justified in concluding that *everything* is both interesting and worthy of study.

When we consider how this prose might be written up as a deduction, we find a small problem similar to one encountered earlier. We are in effect inviting the reader of our deduction to pick anything he wants, intending to show him that *it* is both interesting and worthy of study. But to do this we need a way of specifying whatever it is he picked; once again, we must be able to refer to a thing whose identity may be unknown to us, and once again pseudo-names provide us with the device we need. Of course we must be careful to show this without making use of any special facts about the thing picked; only in that case can we be sure that his *particular* choice of individuals is irrelevant to our deduction – that we could have carried it out as well no matter what was picked and so be justified in concluding that all things are both interesting and worthy of study. So we argue:

1. $(x)(Ix)$	Premise	Everything is interesting
2. $(x)(Wx)$	Premise	Everything is worthy of study
3. Iu	1 UI	Then *u* is interesting, where *u* is anything whatever that you care to pick
4. Wu	2 UI	And *u* – that thing you picked, *whatever* it was – is worthy of study, too
5. Iu & Wu	3, 4 Conj	So of course it's both interesting and worthy of study
6. $(x)(Ix$ & $Wx)$		Since you might have picked anything at all and we've been able to show that no matter what it was it's both interesting and worthy of study, we are justified in concluding that *everything* is both. (If you think we might have *missed* something, go back and imagine you'd picked *it* from the start – our deduction works no matter *which* individual you began with.)

The rule **UG** requires a fairly complicated formulation—after all, it can permit us to pass with justification from expressions like (5) to those like (6) only when the former was obtained in so totally general a manner as to make it evident that the pseudo-name occurring in it may be taken to stand for anything whatsoever.

Universal Generalization (UG)

Where β is a pseudo-name (*not* a constant), one may in a deduction infer from a previously obtained sentence

$\ldots\beta\ldots$

any corresponding universally quantified sentence

$(\mu)(\ldots\mu\ldots)$

provided that: i. β does not occur in any **CP** assumption that has not been closed off;
  ii. β does not occur in any preceding line obtained by **EI**;
  iii. μ is a variable not already occurring in $\ldots\beta\ldots$;
  iv. $\ldots\mu\ldots$ results from $\ldots\beta\ldots$ by replacing each occurrence of β with μ (making *no* other changes).[6]

Our discussion should have made it clear why applying **UG** to a constant is impermissible; from the fact that a particular individual satisfies a certain condition it surely does not follow that everything does. Provisos (i) and (ii) spell out the conditions under which our method of proof will be general enough to permit passage from a sentence in which a pseudo-name occurs to a corresponding universally quantified conclusion. Without (i) we could argue

1. *Fu*	Assumption
2. *(y)(Fy)*	1 UG (mistake, violating proviso (i))
3. *Fu ⊃ (y)(Fy)*	1–2 CP
4. *(x)[Fx ⊃ (y)(Fy)]*	3 UG

thus "proving" something that clearly cannot legitimately be established: that each thing is such that if it's a father then everything is. The **UG** at line (2) is clearly indefensible; '*u*' cannot, at that point, be construed as standing for any old thing anyone might have picked—it is, at that point,

---

[6]As indicated in footnote 2 above, we will in practice only apply **UG** to the pseudo-names '*u*', '*u'*', '*u''*', and so on. Since only '*v*', '*v'*', and so on, are being used in applications of **EI**, then, proviso (ii) is less susceptible to violation than it would otherwise be.

being assumed to stand for one of those things that's a father, and obviously *assuming* that *u* is a father does not justify *concluding* that everything is like *u* in that respect.

Without (ii) we would be permitted to argue from the true premise 'There are fathers' to the false conclusion 'Everything is a father':

1. $(\exists x)(Fx)$ Premise

2. Fv 1 EI

3. $(x)(Fx)$ 2 UG (mistake, violating proviso (ii))

The **UG** at line (3) is obviously in error. The '*v*' in line (2) cannot be construed as standing for anything anyone might have cared to pick; rather, it there stands for some *particular father* we chose at line (2) to consider.

Now if that were the whole story (ii) could be recast merely to block **UG** when the pseudo-name involved came into the deduction by an earlier application of **EI**. But that would not be sufficient, since we could then deduce from the presumably true premise 'Everyone likes someone' the false conclusion 'Someone is liked by everyone':

1. $(x)(\exists y)(Lxy)$ Premise

2. $(\exists y)(Luy)$ 1 UI

3. Luv 2 EI

4. $(x)(Lxv)$ 3 UG (mistake, violating proviso (ii))

5. $(\exists y)(x)(Lxy)$ 4 EG

Even though it was '*v*' that was obtained by **EI** at line (3), '*u*' *occurs* in (3) and that is sufficient, according to proviso (ii), to block the **UG** at (4). To see intuitively why the **UG** at (4) cannot be defended but must be blocked, think of the argument as taking something like the following course:[7]

1. For any person *x* you pick, I can find a person *y* that *x* likes.

2. Oh, you pick person *u*. OK, I can find a person *y* that *u* likes.

3. Here he is, *v* — the person *u* you picked likes this person *v*.

4. So *whoever* you might have picked, he likes *v* and so consequently —

YOU: Wait a minute, here! Sure, after you found I'd picked *u* you were able to find that person *v* whom he likes. But suppose I'd picked person *u'*

---

[7]See J. L. Mackie, "The Rules of Natural Deduction," *Analysis*, vol. 19 (1958), pp. 27–35.

instead of *u* at the beginning. Of course you could have found *some-one u'* likes, premise one assures us of that; but there's no reason to suppose that the person *u'* likes is going to be the *same* person (*v*) that you found out *u* likes.

And of course, I have no reply.

The reason for the restriction in proviso (iii) should be both obvious and familiar; we must not allow such inferences as

$Fua \supset (\exists x)(Fxa)$
$\therefore (x)[Fxa \supset (\exists x)(Fx\cancel{x})]$

for the conclusion would be ambiguous and it would be impossible to tell which of the two quantifiers bind which occurrences of '*x*' in its consequent. Proviso (iv), however, involves a subtlety that is not troublesome in the other rules. If we were permitted to replace just *some* occurrences of the pseudo-name with variables when applying **UG** then we could "prove" that for *any* two things whatever, if the first is a father then so is the second:

→ 1. *Fu*	Assumption	
2. ~~*Fu*	1 DN	
3. *Fu*	2 DN	
4. *Fu* $\supset$ *Fu*	1–3 CP	
5. (*y*)(*Fu* $\supset$ *Fy*)	4 UG (mistake, violating proviso (iv))	
6. (*x*)(*y*)(*Fx* $\supset$ *Fy*)	5 UG	

The best we could get at line (5) would be

5′. (*y*)(*Fy* $\supset$ *Fy*)   4 UG,

replacing *all* occurrences of '*u*' in (4) with '*y*' when doing the **UG**. The situation at line (4) may be described as follows: at that point our reader has picked something arbitrarily — we've used '*u*' to refer to it — and we have succeeded so far in showing that no matter what was picked, if it's a father then it's a father. Now from that it follows that each thing is such that if it's a father then it's a father, that is (5′) follows; but it does *not* follow, and we clearly have not shown, that no matter what was picked, everything is such that it's a father if the thing picked was!

EXERCISES

Using all four of the quantification rules when necessary, deduce the conclusions of the following arguments from their premises:

1. $(x)(Ax \supset Bx)$
   $(x)(Bx \supset Cx)$
   $\therefore (x)(Ax \supset Cx)$

2. $(x)(Dx \supset Ex)$
   $(y)(Ey \supset Fy)$
   $\therefore (z)(Dz \supset Fz)$

3. $(x)[Ax \supset (Bx \ \& \ Cx)]$
   $\sim Bd$
   $\therefore \sim Ad$

4. $(x)[Dx \supset (Ex \ \& \ Fx)]$
   $(\exists x)(\sim Fx)$
   $\therefore (\exists x)(\sim Dx)$

5. $(x)(Mx \supset \sim Lx)$
   $(\exists x)(Nx \ \& \ Lx)$
   $\therefore (\exists x)(Nx \ \& \ \sim Mx)$

6. $(x)[(Bx \ v \ Cx) \supset Dx]$
   $(\exists y)(\sim Cy \ v \ \sim By)$
   $(\exists z)[\sim(Ez \ v \ \sim Cz)]$
   $\therefore (\exists x)(Dx)$

7. $(\exists x)[Ax \ \& \ (y)(By \supset Rxy)]$
   $(x)[Ax \supset (y)(Sy \supset \sim Rxy)]$
   $\therefore (x)(Bx \supset \sim Sx)$

8. All gods are immortal. All gods are moral. Zeus is immoral and Socrates is mortal. Therefore neither Socrates nor Zeus is a god. (Gx-x is a god; Mx-x is mortal; Rx-x is moral; d-Zeus; s-Socrates.)

9. All multiples of six are multiples of two and also multiples of three. All multiples of two are even. Therefore all multiples of six are even. (Mxy-x is a multiple of y; f-six; b-two; c-three; Ex-x is even.)

10. All even numbers are multiples of either six or two. All multiples of six are multiples of two. Therefore all even numbers are multiples of two. (Same notation as (9).)

11. Every triangle is either scalene or isosceles. Isosceles and equilateral triangles are symmetrical. Therefore all triangles that are not scalene are symmetrical. (Tx-x is a triangle; Sx-x is scalene; Ix-x is isosceles; Ex-x is equilateral; Rx-x is symmetrical.)

12. All moths are Lepidoptera. All insects are arthropods. So if all Lepidoptera are insects then all moths are arthropods. (Mx-x is a moth; Lx-x is a Lepidoptera; Ix-x is an insect; Ax-x is an arthropod.)

13. There is at least one being than whom none is greater. Therefore for each being there exists at least one being than which it is not greater. (Gxy-x is greater than y.)

In Section 5.1 we noticed that where μ is any variable any two sentences of the forms

$\sim(\exists\mu)(\ldots\mu\ldots)$

and

$(\mu)(\sim(\ldots\mu\ldots))$

will be equivalent, as will any two sentences of the forms

$\sim(\mu)(\ldots\mu\ldots)$

and

$(\exists\mu)(\sim(\ldots\mu\ldots))$.

Accordingly, we add to our system the associated rule of replacement:

Quantifier Negation (QN)
$\sim(\exists\mu)(\ldots\mu\ldots) :: (\mu)(\sim(\ldots\mu\ldots))$
$\sim(\mu)(\ldots\mu\ldots) :: (\exists\mu)(\sim(\ldots\mu\ldots))$

We *need* not add this rule, for it is redundant and we won't be able to do anything with it that we could not have done more laboriously without it. But it will prove immensely *convenient*.

Without **QN**, for example, the deduction of '$(x)(\sim Fx)$' from '$\sim(\exists x)(Fx)$' requires a sneak attack. Our premise begins with a negation sign, *not* with a quantifier, so of course it is not possible to instantiate it. Instead we have to try something like this:

1. $\sim(\exists x)(Fx)$	Premise	DEDUCE: $(x)(\sim Fx)$
2. Fu	Assumption	
3. $(\exists x)(Fx)$	2 EG	
4. $Fu \supset (\exists x)(Fx)$	2–3 CP	
5. $\sim Fu$	1, 4 MT	
6. $(x)(\sim Fx)$	5 UG	

(notice that the **UG** at line (6) is legitimate – '*u*' does not occur in any line of the deduction obtained by **EI**, and although '*u*' does occur in the assumption made at line (2) that assumption was closed off at line (4) before **UG** was applied). But *with* **QN**, the deduction is a one-step affair:

1. $\sim(\exists x)(Fx)$ Premise   DEDUCE: $(x)(\sim Fx)$

2. $(x)(\sim Fx)$ 1 QN

The usefulness of **QN** will be even more obvious if the reader compares his work on an earlier exercise with the deduction we can now construct:

1. $\sim(\exists x)(Px \,\&\, \sim Wx)$ Premise

2. Pa Premise   DEDUCE: Wa

3. $(x)[\sim(Px \,\&\, \sim Wx)]$ 1 QN

4. $\sim(Pa \,\&\, \sim Wa)$ 3 UI

5. $\sim Pa \lor \sim\sim Wa$ 4 De M

6. $\sim\sim Pa$ 2 DN

7. $\sim\sim Wa$ 5, 6 DS

8. Wa 7 DN

The reader should also notice that **QN**, unlike our other quantification rules, is a rule of *replacement*. With **EI, EG, UI,** and **UG** we can operate only on entire lines of deductions, but with **QN** we can tinker with their parts:

1. $(x)(Fx) \supset (\exists x)(Fx)$ Premise   DEDUCE: $(x)(\sim Fx) \supset (\exists x)(\sim Fx)$

2. $\sim(\exists x)(Fx) \supset \sim(x)(Fx)$ 1 Trans

3. $(x)(\sim Fx) \supset \sim(x)(Fx)$ 2 QN

4. $(x)(\sim Fx) \supset (\exists x)(\sim Fx)$ 3 QN

EXERCISES

Deduce the conclusions of each of the following arguments from their premises:

1. $\sim(x)(Fx \lor Gx)$
      $\therefore (\exists y)(\sim Fy)$

2. $\sim(\exists x)(Fx \lor Gx)$
   $\therefore (x)(\sim Fx) \mathbin{\&} (y)(\sim Gy)$
3. $(x)(\sim Fx)$
   $(\exists y)(Gy) \supset (\exists x)(Fx)$
   $\therefore (y)(\sim Gy)$
4. $(\exists x)[\sim(Fx \mathbin{\&} Gx)]$
   $\sim(x)(Fx \mathbin{\&} Gx) \supset \sim(z)(Hz)$
   $(\exists z)(\sim Hz) \supset (y)(Gy)$
   $\therefore \sim(\exists y)(\sim Gy)$
5. $\sim(\exists x)(\sim Ix)$
   $\sim(\exists x)(\sim Wx)$
   $\therefore \sim(\exists y)(\sim Iy \lor \sim Wy)$

6.2 Strategems

Skill in discovering deductions goes hand in hand with a thorough familiarity with the workings of our rules, and both come with practice. Skill in developing deductions *in a reasonable amount of time* calls for more: ingenuity sometimes, along with a store of strategies and devices to draw on and a firm resolve to avoid beginnings that seem unlikely to lead anywhere.

Some of the arguments we look at have existentially quantified conclusions. In dealing with such arguments the following general rule of thumb is often helpful:

When the conclusion of an argument is existentially quantified, instantiate everything you can; then sort through the resulting lines for something from which the conclusion can be obtained by **EG**.

For example, suppose we're given the argument:

1. $(\exists x)(Hx)$ Premise
2. $(\exists x)\{Qx \,\&\, (y)[Hy \supset (Rx \,\&\, Hx)]\}$ Premise   DEDUCE: $(\exists x)(Hx \,\&\, Qx)$

How are we to obtain '$(\exists x)(Hx \,\&\, Qx)$'? It is barely conceivable that we will somehow deduce something like '$Ab \supset (\exists x)(Hx \,\&\, Qx)$' at line (22) and '$Ab$' at line (37) and so get '$(\exists x)(Hx \,\&\, Qx)$' by **MP**. But it seems far more likely that we will, instead, deduce something like '$Hv \,\&\, Qv$' or '$Hv' \,\&\, Qv''$', and then infer our conclusion by **EG**. There seems no better way to begin moving in that direction than to instantiate our two premises by **EI**, being careful to use different letters for each so as to avoid running afoul of proviso (i) on **EI**:

3. Hv 1 EI
4. $Qv' \,\&\, (y)[Hy \supset (Rv' \,\&\, Hv')]$ 2 EI

These moves suggest at least a few more:

5. $(y)[Hy \supset (Rv' \,\&\, Hv')] \,\&\, Qv'$ 4 Com
6. $(y)[Hy \supset (Rv' \,\&\, Hv')]$ 5 Simp

And obviously (6) can be used to accumulate even more information about v:

7. $Hv \supset (Rv' \,\&\, Hv')$ 6 UI

Glancing again at our conclusion, we see that obtaining it is just a matter of sorting through (3)–(7) until we get '*Hv'* & *Qv''*':

8.	*Rv'* & *Hv'*	3, 7 MP
9.	*Hv'* & *Rv'*	8 Com
10.	*Hv'*	9 Simp
11.	*Qv'*	4 Simp
12.	*Hv'* & *Qv'*	10, 11 Conj
13.	(∃*x*)(*Hx* & *Qx*)	12 EG

Another timesaving rule of thumb suggests itself so readily that the reader has perhaps already discovered it for himself:

*When instantiating premises, do all the **EI**'s first.*

When, for example, we approach the argument

1.	(*x*)[(*Bx* v *Cx*) ⊃ *Dx*]	Premise	
2.	(∃*z*)[~(*Ez* v ~*Cz*)]	Premise	DEDUCE: (∃*x*)(*Dx*)

we first notice the existentially quantified conclusion and see that '*Dv*' would give it to us in one more step. Turning to the business of instantiating premises, we should begin with the **EI** on (2)

3.	~(*Ev* v ~*Cv*)	2 EI

intending to use **UI** next on (1) in order to accumulate more information about *v*:

4.	(*Bv* v *Cv*) ⊃ *Dv*	1 UI

We can then find '*Dv*' and finish:

5.	~*Ev* & ~~*Cv*	3 De M
6.	~*Ev* & *Cv*	5 DN

7. Cv & $\sim Ev$ 6 Com

8. Cv 7 Simp

9. Cv v Bv 8 Add

10. Bv v Cv 9 Com

11. Dv 10, 4 MP

12. $(\exists x)(Dx)$ 11 EG

Had we made the *strategic* mistake of beginning with **UI**

3. $(Bv$ v $Cv) \supset Dv$ 1 UI

what would we do? We could *not* obtain '$\sim(Ev$ v $\sim Cv)$' from (2) by **EI** now because proviso (i) on **EI** blocks use of the letter 'v' (which here already occurs in (3)). So we have to pick a *new* letter for the **EI**:

4. $\sim(Ev'$ v $\sim Cv')$ 2 EI

and what good has the **UI** at (3) done us? To be able to get anywhere with (4) it is v' we need to know about, not v, and so we must **UI** (1) again, to 'v'' this time:

5. $(Bv'$ v $Cv') \supset Dv'$ 1 UI

Notice that (3) is totally useless to us.

When the conclusion of an argument is existentially quantified, then, we begin by instantiating everything we can, being careful to do the **EI**'s first; we then sort through the resulting lines, looking for something from which our conclusion can be inferred by **EG**. But what do we do when our conclusion is *universally* quantified? Most often, in such cases, the conclusion will be of the form

$(\mu)(\ldots\mu\ldots \supset ---\mu---)$

and so conditional proof suggests itself; if we assume $\ldots u \ldots$ (say) and deduce $---u---$, **CP** will give us $\ldots u \ldots \supset ---u---$ and our conclusion will follow by **UG**:

...u...   Assumption
    .
    .
    .
---u---

...u... ⊃ ---u---   CP

(μ)(...μ... ⊃ ---μ---)   UG

Consider, for example, the argument

1. (x)[Fx ⊃ (Gx & Hx)]   Premise
2. (x)(Gx ⊃ Ix)   Premise   DEDUCE: (x)(Fx ⊃ Ix)

To end with '(x)(Fx ⊃ Ix)' we aim first for 'Fu ⊃ Iu', intending to obtain the former from the latter by **UG**; and since 'Fu ⊃ Iu' is a conditional we try to establish it by **CP**, assuming

3. Fu   Assumption

and trying to deduce 'Iu'. To make some use of our assumption it is obvious that we should next instantiate our first premise:

4. Fu ⊃ (Gu & Hu)   1 UI
5. Gu & Hu   3, 4 MP

And now it is clear that **UI** on (2) will quickly lead us to our goal:

6. Gu ⊃ Iu   2 UI
7. Gu   5 Simp
8. Iu   6, 7 MP

Coming out of the conditional proof we have

9. Fu ⊃ Iu   3–8 CP

and **UG** now gives us our conclusion:

10. $(x)(Fx \supset Ix)$ 9 UG

Of course **UG** cannot be applied to a pseudo-name occurring in an assumption *still operative*; but the assumption (3) in which '*u*' occurs here has been closed off at line (9) so the **UG** at (10) is perfectly all right.

Notice how important it is to begin with the assumption '*Fu*' rather than

3. $(x)(Fx)$ Assumption

Had we begun with this assumption, we could have proceeded

4. Fu 3 UI

5. $Fu \supset (Gu \,\&\, Hu)$ 1 UI

6. $Gu \,\&\, Hu$ 4, 5 MP

7. $Gu \supset Iu$ 2 UI

8. Gu 6 Simp

9. Iu 7, 8 MP

but now we would be stymied. If we give up the assumption at this point, to get

10. $(x)(Fx) \supset Iu$ 3–9 CP

we have *no* way of obtaining the desired '$(x)(Fx \supset Ix)$' from (10). The best we can do is apply **UG** to get

11. $(y)[(x)(Fx) \supset Iy]$ 10 UG

And if we try the **UG** on (9) before applying **CP** we can only finish:

10. $(x)(Ix)$ 9 UG

11. $(x)(Fx) \supset (x)(Ix)$ 3–10 CP

We might formulate the general strategy involved here this way:

When the conclusion of an argument is of the form $(\mu)(\ldots\mu\ldots \supset ---\mu---)$, *try assuming* $\ldots\beta\ldots$ *and deducing* $---\beta---$, *intending to use* **CP** *followed by* **UG**.

At the same time, we must notice that there are many arguments which will *not* yield to this *precise* line of attack but only to a simple but easily overlooked modification of it.

Let us confront the innocent-looking argument:

1. $(x)(y)(Exy \supset Eyx)$ Premise

2. $(x)(y)(Exy \supset Exx)$ Premise   DEDUCE: $(x)[(\exists y)(Eyx) \supset Exx]$

Since the conclusion is universally quantified, we naturally think of establishing it by first deducing '$(\exists y)(Eyu) \supset Euu$' and then using **UG**. And since '$(\exists y)(Eyu) \supset Euu$' is a conditional, our first thought is to try to establish it by assuming its antecedent and going for the consequent (this is the step in our thinking that will later have to be modified):

   3. $(\exists y)(Eyu)$ Assumption

The obvious thing to do next is to use **EI** on (3), being careful to use a new pseudo-name, and then look for some help from (1):

4. Evu 3 EI

5. $(y)(Evy \supset Eyv)$ 1 UI

6. $Evu \supset Euv$ 5 UI

7. Euv 4, 6 MP

Things seem to be going well as we turn to (2).

8. $(y)(Euy \supset Euu)$ 2 UI

9. $Euv \supset Euu$ 8 UI

10. Euu 7, 9 MP

11. $(\exists y)(Eyu) \supset Euu$ 3–10 CP

All we need is that final **UG**...

12. $(x)[(\exists y)(Eyx) \supset Exx]$ 11 UG MISTAKE.

The **UG** is *not* permitted! For 'u' occurs in a line — namely (4) — that was obtained by **EI**; and proviso (ii) on **UG** accordingly *blocks* any later attempt to **UG** on 'u'. There does not seem to be any easy way out. Without applying **EI** to (3) it is difficult to see how (10) can be obtained; and if we do apply **EI** to (3) we cannot get (12).

Now in fact (10) can be obtained by indirect proof without using **EI** on (3), but there is a simpler way to carry out the deduction. Look again at the premises, and at the conclusion. The conclusion is universally quantified, and we would like to establish it by first deducing '$(\exists y)(Eyu) \supset Euu$' and then using **UG**. Our *first* thought was to establish the latter by assuming '$(\exists y)(Eyu)$' and deducing 'Euu' from that assumption; unfortunately, in the process of carrying out that conditional proof we made a move that blocked the **UG** we hoped to employ at the end. But there is another way to establish '$(\exists y)(Eyu) \supset Euu$': we can assume the *negation* of its consequent and try to deduce the *negation* of its antecedent; **CP** will then give us '$\sim Euu \supset \sim(\exists y)(Eyu)$' and **Trans** will then turn that into the conditional we want:

1. $(x)(y)(Exy \supset Eyx)$	Premise	
2. $(x)(y)(Exy \supset Exx)$	Premise	DEDUCE: $(x)[(\exists y)(Eyx) \supset Exx]$
3. $\sim Euu$	Assumption	
4. $(y)(Euy \supset Euu)$	2 UI	
5. $Euu' \supset Euu$	4 UI	
6. $\sim Euu'$	3, 5 MT	
7. $(y)(Eu'y \supset Eyu')$	1 UI	
8. $Eu'u \supset Euu'$	7 UI	
9. $\sim Eu'u$	6, 8 MT	
10. $(y)(\sim Eyu)$	9 UG	
11. $\sim(\exists y)(Eyu)$	10 QN	
12. $\sim Euu \supset \sim(\exists y)(Eyu)$	3–11 CP	
13. $(\exists y)(Eyu) \supset Euu$	12 Trans	
14. $(x)[(\exists y)(Eyx) \supset Exx]$	13 UG	

There is nothing suspect about the **UG** at line (14) now—**EI** was never employed in the deduction. Notice, too, that the **UG** at line (10) is perfectly correct—the letter '*u*'' does not occur in the assumption (3).

We might now amend our earlier rule of thumb for dealing with arguments whose conclusions are universally quantified conditionals:

When the conclusion of an argument is of the form $(\mu)(\dots\mu\dots \supset ---\mu---)$, *try assuming* $\dots\beta\dots$ *and deducing* $---\beta---$, *intending to use **CP** followed by* **UG**. *ALTERNATELY, TRY ASSUMING* $\sim(---\beta---)$ *AND DEDUCING* $\sim(\dots\beta\dots)$, *INTENDING TO USE **CP** FOLLOWED BY **TRANS** AND THEN **UG**.*

EXERCISES

I. Without actually beginning any deductions, explain what your *strategy* would be in dealing with each of the following arguments. Then construct a deduction for each.

1. $(x)[(Ax \ \& \ Bx) \supset Cx]$
    $(x)(Dx \supset Bx)$
    $(\exists x)(Ax \ \& \ Dx)$
    $\therefore (\exists x)(Cx \ \& \ Dx)$

2. $(x)[Hx \supset (Bx \equiv \sim Tx)]$
    $(\exists x)(Hx \ \& \sim Bx)$
    $(\exists x)(Hx \ \& \sim Tx)$
    $\therefore (\exists x)(Hx \ \& \ Tx)$

3. $(x)[(\exists y)(Rxy) \supset (\exists z)(\sim Wz)]$
    $(\exists y)(\exists z)(Ryz \ \& \ Hz)$
    $(x)(\sim Hx \supset Wx)$
    $\therefore (\exists z)(\sim Wz \ \& \ Hz)$

4. $(\exists x)(Ux) \supset (y)(z)(Syz \supset Uy)$
    $(\exists x)(\exists y)[(Px \ \& \ Qx) \ \& \ (Ry \ \& \ Syx)]$
    $(x)(Qx \supset Ux)$
    $\therefore (\exists x)(Rx \ \& \ Ux)$

II. As in Exercise I, explain for each of the following arguments what strategy you would employ in deducing its conclusion from its premises; then construct deductions.

1. $(x)[Ax \supset (Bx \lor Cx)]$
    $(x)[(Bx \lor Cx) \supset Dx]$
    $\therefore (x)[(Ax \ \& \sim Bx) \supset (Dx \ \& \ Cx)]$

2. $(x)(y)[(Hx \ \& \ Lxy) \supset Lya]$
    $(x)(y)(Lxa \supset Lxy)$
    $\therefore (x)(y)[(Hx \ \& \ Lxy) \supset Lyx]$

3. $(x)(Rx \supset Fx)$
   $(x)[(Fx \& Gx) \supset Hx]$
   $\therefore (x)(Fx \supset Gx) \supset (y)(Ry \supset Hy)$

III. Deduce the conclusion of each of the following arguments from its premises; be prepared to combine and modify strategies already used, and to invent some new ones of your own.

1. $(\exists x)(Px \& Lxa)$
   $(y)(Py \supset Lay)$
   $(x)(y)[(Lxa \& Lay) \supset Lxy]$
   $\therefore (\exists x)[Px \& (y)(Py \supset Lxy)]$

2. $(x)(y)(Sxy \supset Syx)$
   $(x)(Sxd \supset \sim Tbx)$
   $(x)(y)(Rax \supset Txy)$
   $\therefore Rab \supset \sim Sdc$

3. $(\exists x)(Fx) \supset (y)(Ay \supset By)$
   $(\exists x)(Gx) \supset (y)(Cy \supset \sim By)$
   $\therefore (\exists x)(Gx \& Fx) \supset (y)(Cy \supset \sim Ay)$

4. $(x)(Cx \supset Fx)$
   $\therefore (x)[(\exists y)(Cy \& Dxy) \supset (\exists z)(Fz \& Dxz)]$

5. $(x)(\exists y)(Lxy)$
   $\therefore (y)(\exists z)(Lyz)$

IV. Some of the following arguments are valid and some are invalid. Symbolize each. Then deduce the conclusions of the valid ones from their premises and, after studying Section 5.3, show that the rest are invalid.

1. Everyone doubts something or other. Anyone who doubts something can be certain of at least one thing. Therefore each person can be certain of at least one thing. (Px-x is a person; Dxy-x doubts y; Cxy-x can be certain of y.)

2. For each positive integer x, there exists a sentence containing more than x words. Any sentence that contains, for each positive integer x, more than x words is of infinite length. Therefore there exists at least one sentence of infinite length. (Ix-x is a positive integer; Sx-x is a sentence; Cxy-x contains more than y words; Lx-x is of infinite length.)

3. Only a glamor stock would rise faster than International Toothpick during a depression period. A competitor of International Toothpick rose faster than it during May. Therefore if none of International Toothpick's competitors are glamor stocks, May was not a depression period. (Gx-x is a glamor stock; i-International Toothpick; $Rxyz$-x rises faster than y during z; Dx-x is a depression period; Cxy-x is a competitor of y; m-May.)

4. If one thing resembles a second thing, then the second resembles the first. If one thing resembles a second thing and the second resembles a third, then the first thing resembles the third. At least one thing resembles itself. Therefore everything resembles itself. (*Rxy*-*x* resembles *y*.)

5. Every valid argument Jones presented today was unsound. An argument is sound if and only if it is valid and all its premises are true. Therefore if the Ontological Argument is one of the valid arguments Jones presented today, at least one of its premises is untrue. (*Vx*-*x* is valid; *Ax*-*x* is an argument; *j*-Jones; *Pxy*-*x* presented *y* today; *Sx*-*x* is sound; *Rxy*-*x* is a premise of *y*; *Tx*-*x* is true; *a*-the Ontological Argument.)

6. Every spy who was undetected eluded at least one guard. Some spies were armed and so was each guard they eluded. No armed spy was detected. Therefore some guard is a spy only if he was undetected. (*Sx*-*x* is a spy; *Dx*-*x* was detected; *Exy*-*x* eluded *y*; *Gx*-*x* is a guard; *Ax*-*x* was armed.)

7. There is a politician who is despised by all citizens who despise any politician at all. Every citizen despises at least one politician. Therefore there is a politician who is despised by every citizen. (*Px*-*x* is a politician; *Dxy*-*x* despises *y*; *Cx*-*x* is a citizen.)

8. If one lattice is isomorphic to a second, the second is isomorphic to the first. If one lattice is isomorphic to a second and the second isomorphic to a third, then the first is isomorphic to the third. Therefore if any lattice is isomorphic to some lattice then it is isomorphic to itself. (*Lx*-*x* is a lattice; *Ixy*-*x* is isomorphic to *y*.)

9. If each even integer is such that it satisfies Goldbach's hypothesis if all smaller even integers satisfy that hypothesis, then every even integer satisfies Goldbach's hypothesis. Therefore if for each even integer that does not satisfy Goldbach's hypothesis there exists a smaller even integer that does not satisfy it either, then all even integers satisfy Goldbach's hypothesis. (*Ex*-*x* is an even integer; *Sx*-*x* satisfies Goldbach's hypothesis; *Lxy*-*x* is less than *y*.)

10. If for each even integer that does not satisfy Goldbach's hypothesis there exists a smaller even integer that does not satisfy it either, then every even integer satisfies Goldbach's hypothesis. Therefore if each even integer is such that it satisfies Goldbach's hypothesis if all smaller even integers satisfy that hypothesis, then every even integer satisfies Goldbach's hypothesis. (Same notation as (9).)

*11. If I am morally permitted to perform any action, then everyone is morally permitted to perform that action. If something awful occurs if everyone performs a certain action then no one is morally permitted to perform that action. So if something awful occurs if I perform a certain action then I am not morally permitted to perform that action. (Mxy-x is morally permitted to perform y; Ax-x is an action; Hx-x is a person; Wx-x is awful; Ox-x occurs; Pxy-x performs y; i-me.)

*12. Any argument of interest to all philosophers is worthy of everyone's attention. Anything worthy of anyone's attention ought to be considered by him. Therefore if each philosopher is interested in all arguments propounded by those who disagree with him, then any argument propounded by anyone who disagrees with all philosophers ought to be considered by everyone. (Ax-x is an argument; Ixy-x is interested in y; Px-x is a philosopher; Wxy-x is worthy of y's attention; Hx-x is a person; Oxy-x ought to consider y; Rxy-x propounds y; Dxy-x disagrees with y.)

*6.3 Theorems of Quantificational Logic

In Chapter 3 we noticed that our rules of inference and replacement for the sentential calculus could be used not only to deduce conclusions from premises but also, on occasion, to deduce certain sentences from no premises at all. Such sentences were the *theorems* of sentential logic and they corresponded exactly, as we saw in Chapter 4, to the truth-table tautologies. We have since then extended our symbolic apparatus and our rules to cover quantification as well, and have concentrated thus far on establishing validity; but of course quantificational logic has theorems of its own. They correspond to the logical truths discussed in Section 3 of the preceding chapter: the sentences provable in our system of quantificational logic are precisely those sentences that are true (to use the terminology of that chapter) on all interpretations.[8]

The sentence '$(x)(Fx \lor {\sim}Fx)$', for example, is clearly true no matter what predicate expression 'F' is taken to abbreviate and no matter what universe of discourse the variable 'x' is taken to range over. To show that it is a theorem we have only to construct the following very simple proof:

1.	$Fu \lor {\sim}Fu$	E-M I
2.	$(x)(Fx \lor {\sim}Fx)$	1 UG

Most interesting theorems require more involved proofs, of course. Since no premises are available to us in such situations, the first lines of our deductions can only be justified by **E-M I** (as above) or as assumptions for **CP**. CP is especially useful in establishing conditionals like '$(\exists y)(x)(Cyx) \supset (x)(\exists y)(Cyx)$':

1.	$(\exists y)(x)(Cyx)$	Assumption
2.	$(x)(Cvx)$	1 EI
3.	Cvu	2 UI
4.	$(\exists y)(Cyu)$	3 EG
5.	$(x)(\exists y)(Cyx)$	4 UG
6.	$(\exists y)(x)(Cyx) \supset (x)(\exists y)(Cyx)$	1–5 CP

(Notice that the **UG** at line (5) is correct: the pseudo-name to which it is being applied, 'u', occurs neither in the assumption nor in the line obtained by **EI**.)

---

[8]The rather complicated demonstration of this claim for a system of quantificational logic essentially equivalent in deductive power to the present one is carried out in E. Mendelson, *Introduction to Mathematical Logic*, Princeton: Van Nostrand, 1964.

Often more complicated strategies are advantageous. It was noted in Chapter 5 that '$[(x)(Fx) \vee (y)(Gy)] \supset (z)(Fz \vee Gz)$' is a logical truth (though its converse is not); if we try to prove it by assuming its antecedent and proceeding to deduce its consequent, we are immediately faced with a problem:

> 1. $(x)(Fx) \vee (y)(Gy)$ Assumption

(1), as it stands, cannot be instantiated. Before using it we have to do some additional work that does not suggest itself immediately:

>> 2. $(x)(Fx)$ Assumption
   3. Fu 2 UI
   4. $Fu \vee Gu$ 3 Add
   5. $(z)(Fz \vee Gz)$ 4 UG

6. $(x)(Fx) \supset (z)(Fz \vee Gz)$ 2–5 CP

>> 7. $(y)(Gy)$ Assumption
   8. Gu 7 UI
   9. $Gu \vee Fu$ 8 Add
   10. $Fu \vee Gu$ 9 Com
   11. $(z)(Fz \vee Gz)$ 10 UG

12. $(y)(Gy) \supset (z)(Fz \vee Gz)$ 7–11 CP
13. $(z)(Fz \vee Gz) \vee (z)(Fz \vee Gz)$ 1, 6, 12 CD
14. $(z)(Fz \vee Gz) \vee {\sim}(z)(Fz \vee Gz)$ E-M I
15. $[(z)(Fz \vee Gz) \vee (z)(Fz \vee Gz)] \,\&$
    $[(z)(Fz \vee Gz) \vee {\sim}(z)(Fz \vee Gz)]$ 13, 14 Conj
16. $(z)(Fz \vee Gz) \vee [(z)(Fz \vee Gz) \,\&$
    ${\sim}(z)(Fz \vee Gz)]$ 15 Dist
17. $[(z)(Fz \vee Gz) \,\& {\sim}(z)(Fz \vee Gz)] \vee$
    $(z)(Fz \vee Gz)$ 16 Com
18. ${\sim}(z)(Fz \vee Gz) \vee {\sim}{\sim}(z)(Fz \vee Gz)$ E-M I
19. ${\sim}[(z)(Fz \vee Gz) \,\& {\sim}(z)(Fz \vee Gz)]$ 18 De M
20. $(z)(Fz \vee Gz)$ 19, 17 DS

21. $[(x)(Fx) \vee (y)(Gy)] \supset (z)(Fz \vee Gz)$ 1–20 CP

The proof is considerably less messy if we employ a more familiar strategy and instead begin by assuming the negation of the consequent of the theorem we're trying to establish and aim for the negation of its antecedent, finishing off with a **Trans** step:

1.	~(z)(Fz v Gz)	Assumption
2.	(∃z)[~(Fz v Gz)]	1 QN
3.	~(Fv v Gv)	2 EI
4.	~Fv & ~Gv	3 De M
5.	~Fv	4 Simp
6.	(∃x)(~Fx)	5 EG
7.	~(x)(Fx)	6 QN
8.	~Gv	4 Com, Simp
9.	(∃y)(~Gy)	8 EG
10.	~(y)(Gy)	9 QN
11.	~(x)(Fx) & ~(y)(Gy)	7, 10 Conj
12.	~[(x)(Fx) v (y)(Gy)]	11 De M
13.	~(z)(Fz v Gz) ⊃ ~[(x)(Fx) v (y)(Gy)]	1–12 CP
14.	[(x)(Fx) v (y)(Gy)] ⊃ (z)(Fz v Gz)	13 Trans

When the theorem to be established is a biconditional, it is often best to divide the proof into two parts, establishing the two halves of the biconditional separately. To prove '(x)(Fx & Gx) ≡ [(y)(Fy) & (z)(Gz)]', for example, we might begin with:

1.	(x)(Fx & Gx)	Assumption
2.	Fu & Gu	1 UI
3.	Fu	2 Simp
4.	(y)(Fy)	3 UG
5.	Gu	2 Com, Simp
6.	(z)(Gz)	5 UG
7.	(y)(Fy) & (z)(Gz)	4, 6 Conj
8.	(x)(Fx & Gx) ⊃ [(y)(Fy) & (z)(Gz)]	1–7 CP

With that half out of the way, we can proceed in the other direction:

9. $(y)(Fy)$ & $(z)(Gz)$	Assumption
10. $(y)(Fy)$	9 Simp
11. Fu	10 UI
12. $(z)(Gz)$	9 Com, Simp
13. Gu	12 UI
14. Fu & Gu	11, 13 Conj
15. $(x)(Fx$ & $Gx)$	14 UG
16. $[(y)(Fy)$ & $(z)(Gz)] \supset (x)(Fx$ & $Gx)$	9–15 CP

Then it is simply a matter of putting the two sub-conclusions together:

17. $\{(x)(Fx$ & $Gx) \supset [(y)(Fy)$ & $(z)(Gz)]\}$ & $\{[(y)(Fy)$ & $(z)(Gz)] \supset (x)(Fx$ & $Gx)\}$	8, 16 Conj
18. $(x)(Fx$ & $Gx) \equiv [(y)(Fy)$ & $(z)(Gz)]$	17 Equiv.

In showing (18) to be a theorem — and hence a logical truth — we have shown the sentences '$(x)(Fx$ & $Gx)$' and '$(y)(Fy)$ & $(z)(Gz)$' to be *equivalent*. For clearly two sentences whose corresponding biconditional is a theorem must be true under precisely the same conditions. Thus, the deductive system of the present chapter can be used to establish equivalence as well as validity.

Of course the proofs of certain theorems call for a combination of strategies. To establish '$(x)\{\sim [Gx \equiv (Gx$ & $Fx)] \supset \sim Fx\}$' we might argue:

1. Fu	Assumption
2. Gu	Assumption
3. Gu & Fu	2, 1 Conj
4. $Gu \supset (Gu$ & $Fu)$	2–3 CP
5. Gu & Fu	Assumption
6. Gu	5 Simp
7. $(Gu$ & $Fu) \supset Gu$	5–6 CP
8. $[Gu \supset (Gu$ & $Fu)]$ & $[(Gu$ & $Fu) \supset Gu]$	4, 7 Conj
9. $Gu \equiv (Gu$ & $Fu)$	8 Equiv

10. $Fu \supset [Gu \equiv (Gu \ \& \ Fu)]$ 1–9 CP

11. $\sim[Gu \equiv (Gu \ \& \ Fu)] \supset \ \sim Fu$ 10 Trans

12. $(x)\{\sim[Gx \equiv (Gx \ \& \ Fx)] \supset \ \sim Fx\}$ 11 UG

And as a sort of last resort one can always employ an *indirect* proof, assuming the contradictory of the theorem he is trying to prove and deducing a contradiction from that assumption:

1. $(\exists y)(x)(Rxx \ \& \sim Rxy)$ Assumption

2. $(x)(Rxx \ \& \sim Rxv)$ 1 EI

3. $Rvv \ \& \sim Rvv$ 2 UI

4. $(\exists y)(x)(Rxx \ \& \sim Rxy) \supset (Rvv \ \& \sim Rvv)$ 1–3 CP

5. $\sim Rvv \ v \sim\sim Rvv$ E-M I

6. $\sim(Rvv \ \& \sim Rvv)$ 5 De M

7. $\sim(\exists y)(x)(Rxx \ \& \sim Rxy)$ 6, 4 MT

EXERCISES

I. Show that each of the following is a theorem of our system of quantificational logic:
1. $(\exists x)(Fx \ \& \ Gx) \supset [(\exists y)(Fy) \ \& \ (\exists z)(Gz)]$
2. $(x)[(\exists y)(\sim Gxy) \supset \ \sim(z)(Gxz)]$
3. $(\exists x)(Fx \ v \ Gx) \equiv [(\exists y)(Fy) \ v \ (\exists z)(Gz)]$
4. $(x)[Mx \supset (\exists y)(Vy)] \equiv [(\exists x)(Mx) \supset (\exists y)(Vy)]$
5. $(x)[Fx \supset (y)(Gxy)] \equiv (x)(y)(Fx \supset Gxy)$
6. $(x)(Fx \ v \ Ga) \equiv [(y)(Fy) \ v \ Ga]$

II. Prove the following theorems, and then discuss their relationship to our rules **EI, EG, UI,** and **UG**:
1. $(\exists y)[(\exists x)(Fx) \supset Fy]$
2. $(y)[Fy \supset (\exists x)(Fx)]$
3. $(y)[(x)(Fx) \supset Fy]$
4. $(\exists y)[Fy \supset (x)(Fx)]$

III. *Without* using the rule **QN**, prove the following biconditionals:
1. $\sim(\exists x)(Fx) \equiv (y)(\sim Fy)$
2. $\sim(x)(Fx) \equiv (\exists y)(\sim Fy)$

∗6.4 Rules for '='

Certain arguments whose premises or conclusions contain the identity sign, '=', suggest the addition of two final rules. The first allows us, in effect, to introduce the truism 'Each thing is identical with itself' into any deduction, so we call it

Identity Introduction (II)

At any point in any deduction, one may introduce the sentence

$(x)(x = x)$.

Using this rule we can handle the argument 'The numbers 3, 7 and 21 are factors of the number 231; therefore one of the factors of the number 231 is identical with the number 3' along the following lines:

1. *(Fad & Fbd) & Fcd*	Premise	DEDUCE: $(\exists x)(Fxd \ \& \ x = a)$
2. *Fad & Fbd*	1 Simp	
3. *Fad*	2 Simp	
4. $(x)(x = x)$	II	
5. $a = a$	4 UI	
6. *Fad* $\& \ a = a$	3, 5 Conj	
7. $(\exists x)(Fxd \ \& \ x = a)$	6 EG	

But another rule is needed for the argument

No island is a peninsula
Formosa is an island off the coast of China
Formosa and Taiwan are identical
Therefore Taiwan is not a peninsula.

We can begin easily enough:

1. $(x)(Ix \supset \ \sim Px)$	Premise	
2. *If & Ofc*	Premise	
3. $f = t$	Premise	DEDUCE: $\sim Pt$
4. *If*	2 Simp	
5. *If* $\supset \ \sim Pf$	1 UI	
6. $\sim Pf$	4, 5 MP	

But now what? It seems entirely clear (on the intuitive level) that our conclusion follows from (6) and (3) so that we are entitled to move to

7. $\sim Pt$ 6, 3 ?

After all, if Formosa is not a peninsula—as (6) tells us—and Formosa and Taiwan are *identical*—as (3) says—then Taiwan must not be a peninsula either. With slogans about substituting "equals" for "equals" in mind, we are naturally led to formulate the following rule:

Identity Elimination (IE)

Where α is any pseudo-name or constant, and γ is any pseudo-name or constant, one may in a deduction infer from two preceding sentences

   $\ldots\alpha\ldots$

and

   $\alpha = \gamma$

any corresponding sentence

   $\ldots\gamma\ldots$

provided that $\ldots\gamma\ldots$ is just like $\ldots\alpha\ldots$ except for containing γ at one or more places where $\ldots\alpha\ldots$ contains α.

Pseudo-names as well as constants must be allowed if we are to carry out the deduction of 'Any place Bob is visiting this summer is in the Southwest' from 'The Grand Canyon is in the Southwest' and 'If Bob is visiting any place this summer, it is the Grand Canyon'.

1. Sg Premise

2. $(x)[(Px \mathrel{\&} Vbx) \supset g = x]$ Premise   DEDUCE: $(x)[(Px \mathrel{\&} Vbx) \supset Sx]$

   → 3. $Pu \mathrel{\&} Vbu$ Assumption

     4. $(Pu \mathrel{\&} Vbu) \supset g = u$ 2 UI

     5. $g = u$ 3, 4 MP

     6. Su 1, 5 IE

7. $(Pu \mathrel{\&} Vbu) \supset Su$ 3–6 CP

8. $(x)[(Px \mathrel{\&} Vbx) \supset Sx]$ 7 UG

And **IE** permits replacement of only *some* of the occurrences of the one pseudo-name or constant by the other so that we may argue:

1. $f = t$	Premise	DEDUCE: $t = f$
2. $(x)(x = x)$	II	
3. $f = f$	2 UI	
4. $t = f$	3, 1 IE	

The trick here is to use (1) to replace just *one* 'f' in the otherwise useless-looking (3). Obviously the same deduction can be generalized to provide a proof for '$(x)(y)(x = y \supset y = x)$', i.e. a proof that the identity relation is symmetric:

→ 1. $u = u'$	Assumption
2. $(x)(x = x)$	II
3. $u = u$	2 UI
4. $u' = u$	3, 1 IE
5. $u = u' \supset u' = u$	1–4 CP
6. $(y)(u = y \supset y = u)$	5 UG
7. $(x)(y)(x = y \supset y = x)$	6 UG

A modification of this strategy permits one to prove that identity is transitive as well (see Exercise I (1) below), and reflexivity is taken care of directly by **II**.

Since Formosa is an island and China is not, China is of course not identical with Formosa. To draw the latter conclusion from the premise, **IE** can be combined with an indirect approach:

1. $If \ \& \sim Ic$	Premise	DEDUCE: $f \neq c$
→ 2. $f = c$	Assumption	
3. If	1 Simp	
4. Ic	3, 2 IE	
5. $f = c \supset Ic$	2–4 CP	
6. $\sim Ic$	1 Com, Simp	
7. $f \neq c$	5, 6 MT	

(writing '$f \neq c$' for '$\sim f = c$', as usual).

Sentences involving definite descriptions pick up identity signs as they go into symbols and so provide additional examples of **IE** in use. Notice its role, for example, in the following deduction of 'Only a leader of the reform faction will be nominated' from 'The leader of the party is a leader of the reform faction' and 'The nominees will all be leaders of the party':

1. $(\exists x)\{[Lxp \,\&\, (y)(Lyp \supset x = y.]\,\&\, Lxf\}$	Premise	DEDUCE: $(x)(Nx \supset Lxf)$
2. $(x)(Nx \supset Lxp)$	Premise	
3. Nu	Assumption	
4. $Nu \supset Lup$	2 UI	
5. Lup	3, 4 MP	
6. $[Lvp \,\&\, (y)(Lyp \supset v = y)]\,\&\, Lvf$	1 EI	
7. $Lvp \,\&\, (y)(Lyp \supset v = y)$	6 Simp	
8. $(y)(Lyp \supset v = y)$	7 Com, Simp	
9. $Lup \supset v = u$	8 UI	
10. $v = u$	9, 5 MP	
11. Lvf	6 Com, Simp	
12. Luf	11, 10 IE	
13. $Nu \supset Luf$	3–12 CP	
14. $(x)(Nx \supset Lxf)$	13 UG	

The careful reader will have noticed that we have been especially timorous in discussing **IE**. We have shown how it can be used to justify quite naturally the steps required for a number of deductions, but we have not treated this rule with the same measure of confidence and finality as earlier ones — and there is a good reason for this.

IE licenses, in effect, the substitution of one term for another whenever we can establish the corresponding identity sentence. One wonders, then, how **IE** may be justified or formulated so as to block replacement of 'Venus' with 'Hesperus' in

Poets sometimes play on the fact that Venus and Hesperus are one and the same,

on the basis of the true identity sentence

Venus is identical with Hesperus.

Such a replacement yields a disconcerting result:

Poets sometimes play on the fact that Hesperus and Hesperus are one and the same.

The problem of accounting for the invalidity of these arguments while licensing those given earlier has long been a subject of philosophical research.[9] We mention only that the contexts involved in the exercises below are ones in which there is general agreement that application of **IE** is perfectly acceptable. The reader should be on his guard, however, when he is considering application of **IE** in contexts encountered outside textbook examples.

EXERCISES

I. Prove each of the following theorems:
  1. $(x)(y)(z)[(x = y \,\&\, y = z) \supset x = z]$
  2. $(x)(y)(z)[(x = y \,\&\, x = z) \supset z = y]$
  3. $(x)(y)[(Fx \,\&\, \sim Fy) \supset x \neq y]$

II. Deduce the conclusions of each of the following arguments from their premises:
  1. Taiwan is either an island or a peninsula. Formosa is not a peninsula. Formosa and Taiwan are identical. Therefore Taiwan is an island.
  2. Samuel Clemens was a humorist but Samuel Morse was not. Samuel Clemens is identical with Mark Twain. Therefore Samuel Morse is not identical with Mark Twain.
  3. Lewis Carroll is the author of *Alice in Wonderland*. Anyone who authored *Alice in Wonderland* is the sole author of *The Game of Logic*. Lewis Carroll is Charles Dodgson. Therefore Charles Dodgson is the sole author of *The Game of Logic*.
  4. At least one person is Jones' sister. All of Jones's sisters visited him in the hospital. At most one person visited Jones in the hospital. Therefore exactly one person visited Jones in the hospital.
  5. Jones has at least one sister. All of Jones's sisters visited him in hospital. At most one person visited Jones in the hospital. Anyone who visited Jones in the hospital brought flowers. Of course all of Jones' sisters are persons. Therefore Jones's only sister brought flowers.

---

[9]See, for example, W. V. Quine, *From a Logical Point of View*, Cambridge: Harvard Univ. Press, 1953, Chapter VIII("Reference and Modality"); L. Linsky, *Referring*, London: Routledge and Kegan Paul, 1967, Chapters 2–4; and the references cited in footnote 19 of Section 5.4.

6. At least one well-qualified person will be hired. At least one experienced person will be hired. At most one person will be hired. Therefore the person who will be hired is both well-qualified and experienced.

7. $(x)(y)[(Rxy \ \& \ Ryx) \supset x = y]$
$(x)(y)(Rxy \supset Ryx)$
$\therefore (x)[(\exists y)(Rxy \lor Ryx) \supset Rxx]$

8. $(x)(y)[(Rxy \ \& \ Ryx) \supset x = y]$
$(x)(y)(Rxy \supset Ryx)$
$\therefore (x)\{[(\exists y)(Rxy) \lor (\exists z)(Rxz)] \supset Rxx\}$

9. The author of *Waverley* is bald. Scott is the author of *Waverley*. Therefore Scott is bald.

III. From the single premise 'If every member of one set is a member of a second, and every member of the second is a member of the first, then the sets in question are identical' (sometimes called the *Axiom of Extensionality*), deduce each of the following.

1. For any two sets there exists at most one set whose members are precisely those things that are members of both of the others.

2. If there exists a set with no members, then there is exactly one set with no members.

3. If x and y are sets, then x and y are identical if and only if each member of either is a member of the other.

*IV. Show that the sentence

I want a girl just like the girl that married dear old Dad

when translated with the description having *primary* occurrence implies the translation in which it has *secondary* occurrence.

APPENDIX A

NORMAL FORMS AND CIRCUITS

By a **literal** we shall mean any sentence letter or its negation; thus '*A*' and '~*B*' are literals but '~~*A*' and '*A* & *B*' are not. By a **disjunction of literals** we mean any disjunction all of whose disjuncts are literals but in which no literal occurs more than once. In this appendix, internal parentheses for extended disjunctions will be omitted, and hence not only '*A* v ~*B*' but '~*A* v *B* v *C*' are disjunctions of literals. But '*A* v *A*' is not, and neither is '~~*A* v *B* v (*A* & *C*)', since the first and third disjuncts are not literals. We shall likewise omit internal parentheses for extended conjunctions, and shall mean by a **conjunction of literals** any conjunction all of whose conjuncts are literals but in which no literal occurs more than once. Both '*A* & ~*B*' and '~*A* & *B* & *C*' are thus conjunctions of literals, but '*A* & *A*' and '*A* & ~~*B* & (*A* v *C*)' are not. Moreover, '*A* & ~*A*' is a conjunction of literals, though the same *letter* occurs more than once.

We are now in a position to define two concepts of sentential logic that will be of importance to us here.

A truth-functional sentence is in **conjunctive normal form (CNF)** if and only if it is either (i) a literal, (ii) a disjunction of literals, or (iii) a conjunction all of whose conjuncts are either literals or disjunctions of literals but in which no conjunct occurs more than once.[1]

Thus '*A*', '*A* v ~*B*', '*A* & (*C* v *B* v ~*D*)', '(*A* v *B*) & (*B* v ~*C*)', and '*A* & *B*' are in CNF, but '(*C* v *D*) & (*C* v *D*)', '(*A* & *B*) v *D*', and '~~*B* & (*A* v *D*)' are not.

---

[1]The notion of conjunctive normal form employed here differs from that of Chapter 4 because of our present insistence on lack of duplication.

A truth-functional sentence is in **disjunctive normal form (DNF)** if and only if it is either (i) a literal, (ii) a conjunction of literals, or (iii) a disjunction all of whose disjuncts are either literals or conjunctions of literals but in which no disjunct occurs more than once.

Thus, '$\sim A$', 'A & $\sim B$', 'A v $(C$ & B & $\sim D)$', '$(A$ & $B)$ v $(B$ & $\sim C)$', and 'A v $\sim B$' are in DNF, but '$(C$ & $D)$ v $(C$ & $D)$', '$(A$ v $B)$ & D', and '$\sim\sim B$ v $(A$ & $D)$' are not.

Notice then that any literal is in *both* CNF and DNF. Moreover, disjunctions of literals are in DNF as well as CNF, and conjunctions of literals are in CNF as well as DNF. However, '$(A$ & $\sim B)$ v D' is in DNF only, and $(A$ v $\sim B)$ & D' is in CNF only. On the other hand, 'A v $(B$ & $(C$ v $D))$' and '$\sim\sim A$ & $(B$ v $C)$' are in neither form. It should be further noted that any sentence in either of the two forms will have the connective '$\sim$' attaching only to single sentence letters. Thus '$\sim(A$ & $B)$' is neither in CNF nor in DNF, and the same goes for '$\sim(A$ v $B)$'.

Any truth-functional sentence whatever is either *in* CNF or is *equivalent* to (or, by Section 4.2, *interdeducible* with) a sentence in CNF. In the latter case, it may be *reduced* to the CNF sentence by repeated application of the rules of replacement in Chapter 3, together with some additional rules that will simplify our work. That is, we shall employ **DN, Com, Assoc**, as well as expanded versions of **DeM** and **Dist**, to accommodate extended conjunctions and disjunctions:

De Morgan: $\sim(p_1$ & $\ldots$ & $p_n)$ :: $\sim p_1$ v $\ldots$ v $\sim p_n$
$\sim(p_1$ v $\ldots$ & $p_n)$ :: $\sim p_1$ & $\ldots$ & $\sim p_n$

Distribution: p & $(q_1$ v $\ldots$ v $q_n)$ :: $(p$ & $q_1)$ v $\ldots$ v $(p$ & $q_n)$
p v $(q_1$ & $\ldots$ & $q_n)$ :: $(p$ v $q_1)$ & $\ldots$ & $(p$ v $q_n)$

In addition, we shall employ two forms of the **Law of Idempotence** (abbreviated **Idem**), whose legitimacy is easily established by a two-row truth-table.

Idempotence: p :: p v p
p :: p & p

Clearly a conjunction with a sentence of the form 'p v $\sim p$' as a conjunct is equivalent to the other conjunct alone since the whole will be true if and only if that latter conjunct is true, and false otherwise. Similarly, a disjunction with a sentence of the form 'p & $\sim p$' as a disjunct is equivalent to the other disjunct by itself. Thus, a conjunct of the form 'p v $\sim p$' and a disjunct of the form 'p & $\sim p$' are both superfluous and may therefore be eliminated. We may add, then, two more rules, **Tautology Elimination (TE)** and **Contradiction Elimination (CE)**:

TE: $p :: p$ & $(q \vee \sim q)$
CE: $p :: p \vee (q$ & $\sim q)$

We shall also employ the familiar **Impl** and **Equiv** principles for eliminating '$\supset$' and '$\equiv$' respectively.

To reduce a sentence not in CNF to one that is, we proceed in five basic steps (though in simple cases one or more of these may not be necessary). First, use **Impl** and **Equiv** to eliminate any occurrences of '$\supset$' and '$\equiv$'. Second, use **De M** repeatedly to drive all curls that do not attach to single letters as far inside as possible. Third, use **DN** to eliminate excess curls. Fourth, use **Com**, **Assoc**, and **Dist** to obtain (if necessary) conjunctions of disjunctions, and fifth, use **Idem** and **CE** to eliminate any redundancies. As an example, let us consider:

(1) $\sim \{[B$ & $\sim((C$ & $\sim B) \vee \sim D)] \vee (A \supset D)\}$.

At the first step, we eliminate '$\supset$':

$\sim \{[B$ & $\sim((C$ & $\sim B) \vee \sim D)] \vee (\sim A \vee D)\}$.   Impl

Next, we use **De M** to drive curls inside:

$\sim [B$ & $\sim((C$ & $\sim B) \vee \sim D)]$ & $\sim(\sim A \vee D)$ De M
$[\sim B \vee \sim\sim((C$ & $\sim B) \vee \sim D)]$ & $(\sim\sim A$ & $\sim D)$ De M (twice)

The double negative on the lefthand side still does not attach to a single letter, but instead of applying **De M** any further, it is easier to move on to the third stage and apply **DN** wherever we can:

$[\sim B \vee ((C$ & $\sim B) \vee \sim D)]$ & $(A$ & $\sim D)$ DN (twice)

Now at our fourth step we apply **Com** and **Dist** to the right disjunct of the lefthand side (for brevity, we shall combine **Com** with applications of other rules into one step):

$[\sim B \vee ((\sim D \vee C)$ & $(\sim D \vee \sim B))]$ & $(A$ & $\sim D)$ Com, Dist

Clearly another **Dist** move is required on the lefthand side:

[(($\sim B$ v ($\sim D$ v C)) & ($\sim B$ v ($\sim D$ v $\sim B$))] & (A & $\sim D$)   Dist

Note that this entire expression is of the form '(p & q) & (r & s)', and we know by **Assoc** that placement of these parentheses is immaterial. So in accordance with our earlier decision to omit them, we now drop them out, citing **Assoc** as our justification.[2]

($\sim B$ v ($\sim D$ v C)) & ($\sim B$ v ($\sim D$ v $\sim B$)) & A & $\sim D$ Assoc

The inner parentheses of the two disjunctions are also dispensable, and dropping them now gives:

($\sim B$ v $\sim D$ v C) & ($\sim B$ v $\sim D$ v $\sim B$) & A & $\sim D$ Assoc

Finally, the literal '$\sim B$' occurs twice in the disjunction '$\sim B$ v $\sim D$ v $\sim B$'. So **Com** and **Idem** will eliminate one of them, and the result is a sentence in CNF:

(1′) ($\sim B$ v $\sim D$ v C) & ($\sim B$ v $\sim D$) & A & $\sim D$ Com, Idem

(1′) is a conjunction each of whose conjuncts is either a literal or a disjunction of literals. Of course a reduction to CNF is rarely as involved as the above; it was used here so that all five steps could be fully illustrated.

Often, a sentence reduced to CNF can be further reduced to a *simpler* CNF. An elementary example would be:

(2) (A v B) & (A v $\sim B$) & ($\sim A$ v B).

Though (2) is in CNF, it is equivalent to another, less complex sentence in CNF. Let us use **Dist** on the two left conjuncts; we thus obtain

(A v (B & $\sim B$)) & ($\sim A$ v B)   Dist

---

[2]We are obviously using a different version of **Assoc** here than in Chapter 3. There, it was employed to *move about* parentheses, whereas here it is employed to get rid of them. But since they *can* be moved about without affecting truth-conditions, they are dispensable, and **Assoc** is the rule that makes this idea explicit.

By **CE**, the left side reduces to '*A*', giving

A & (∼A v B)   CE

Dist, Com, and **CE** again now yield the simpler CNF.

(A & ∼A) v (A & B)   Dist
(2') A & B   Com, CE

Since it is a conjunction of distinct literals, clearly (2') is in CNF.

The foregoing considerations apply equally to DNF. Any truth-functional sentence is either *in* DNF or else is *equivalent* to a sentence in DNF, and moreover the same replacement rules and reduction process may be used to find, for any sentence not in DNF, one that is. Consider for example:

(3) (A & ∼B) ⊃ [C ≡ (C v D)]

Proceeding as we did before, we derive successively:

∼(A & ∼B) v [C ≡ (C v D)]   Impl
∼(A & ∼B) v [(C & (C v D)) v (∼C & ∼(C v D))]   Equiv
(∼A v ∼∼B) v [(C & (C v D)) v (∼C & (∼C & ∼D))]   De M (twice)
(∼A v B) v [(C & (C v D)) v (∼C & (∼C & ∼D))]   DN
(∼A v B) v [((C & C) v (C & D)) v (∼C & (∼C & ∼D))]   Dist
∼A v B v (C & C) v (C & D) v (∼C & ∼C & ∼D)   Assoc
(3') ∼A v B v C v (C & D) v (∼C & ∼D)   Idem (twice)

Clearly (3') is in DNF, since each disjunct is either a literal or else a conjunction of literals. However, a simpler DNF is obtainable. We commute the third and fourth disjuncts and then distribute.

∼A v B v (C & D) v ((C v ∼C) & (C v ∼D))   Com, Dist
∼A v B v (C & D) v (C v ∼D)   Com, TE
∼A v B v C v ∼D v (C & D)   Com, Assoc
∼A v B v C v ((∼D v C) & (∼D v D))   Dist
∼A v B v C v (∼D v C)   Com, TE
∼A v B v C v C v ∼D   Com, Assoc
(3'') ∼A v B v C v ∼D   Idem

Normal forms have many uses. In Chapter 4 a similar notion of CNF is employed to establish some theoretical results concerning the concepts and methods of the preceding two chapters. In addition, certain logical features of a complex truth-functional sentence are often more easily discerned if it is reduced to an equivalent CNF or DNF—for example, whether it is a tautology, contradiction, or neither. In general, the truth-conditions of a sentence are displayed more perspicuously if it is in normal form than otherwise. However, one of the most interesting applications concerns electrical circuits, to which we shall turn momentarily.

EXERCISES

I. Put each of the following into *either* CNF *or* DNF—whichever seems most readily obtainable. Then see if there is a *simpler* normal form.
   1. $(A \,\&\, \sim B) \supset (C \vee \sim A)$
   2. $\sim A \,\&\, (\sim A \vee B)$
   3. $\sim [D \vee (B \,\&\, \sim C)]$
   4. $D \equiv \sim (E \,\&\, F)$
   5. $[(A \supset B) \supset A] \supset A$
   6. $\sim [A \supset (B \,\&\, A)]$
   7. $(A \vee B) \supset [(A \supset (B \,\&\, C)) \supset (C \,\&\, \sim A)]$

We now examine some special cases of normal forms.

A truth-functional sentence is a **conjunctive Boolean expansion** if and only if it is in conjunctive normal form and, moreover, every sentence letter occurring (with or without an accompanying negation sign) in *any* of its conjuncts occurs in *all* its conjuncts (perhaps occurring negated in some of these conjuncts and unnegated in the others).[3]

Thus, '$A \vee \sim C$', '$(\sim A \vee \sim B) \,\&\, (\sim A \vee B) \,\&\, (A \vee B)$', and '$(E \vee \sim F \vee D) \,\&\, (\sim E \vee F \vee D)$' are all conjunctive Boolean expansions (hereafter CBE). Every CBE is in CNF, though of course the reverse does not hold. Rather than attempt a definition as tortuous as the above for *disjunctive* expansions, we shall try another approach—one that brings out the complementary nature of the two kinds of expansion.

A truth-functional sentence is a **disjunctive Boolean expansion** if and only if it can be obtained from a CBE by replacing each '&' with 'v' and vice versa.

---

[3]Sentences with only one component letter are a limiting case: 'A', '$\sim B$' and '$\sim C \,\&\, C$' count as CBE's.

Correspondingly, then, 'A & $\sim C$', '($\sim A$ & $\sim B$) ∨ ($\sim A$ & B) ∨ (A & B)', and '(E & $\sim F$ & D) ∨ ($\sim E$ & F & D)' are all disjunctive Boolean expansions (DBE). Clearly a DBE is in DNF.

Every truth-functional sentence either is a CBE (a DBE) or else is equivalent to a CBE (and a DBE) containing the same number of letters. Reduction of a sentence to a Boolean expansion can be achieved using the same replacement rules used earlier for normal forms. However, there is another way of obtaining a Boolean expansion. One may go directly to the truth-table for the sentence, and produce its expansion from the assignment of **T**'s and **F**'s in its column. Consider for example

A	B	$A \equiv B$
T	T	T
T	F	F
F	T	F
F	F	T

A	B	C	(A ∨ B) ⊃ (A & $\sim C$)
T	T	T	F
T	T	F	T
T	F	T	F
T	F	F	T
F	T	T	F
F	T	F	F
F	F	T	T
F	F	F	T

Now a glance at the table tells us that '$A \equiv B$' is true in just two cases: when 'A' and 'B' are both true (that is, 'A & B' is true) or when 'A' and 'B' are both false (that is, when '$\sim A$ & $\sim B$' is true). Thus '$A \equiv B$' is true under just the same conditions as the disjunction '(A & B) ∨ ($\sim A$ & $\sim B$)', which is a DBE. To find a DBE, therefore, we need only to look at those rows in which a **T** is assigned; for each such row we conjoin the component letters, negating those to which an **F** is assigned in that row, and disjoin all the conjunctions thus formed. If just one row has a **T**, we of course will not have a disjunction but instead a conjunction of literals — any sentence having a **T** in just the third row of the lefthand table may be reexpressed as the DBE '$\sim A$ & B'. But where two or more rows have a **T**, the DBE will have exactly as many disjuncts as there are rows assigned **T**.

Turning now to '(A ∨ B) ⊃ (A & $\sim C$)', it will be true just in case either 'A & B & $\sim C$' is true (second row) or 'A & $\sim B$ & $\sim C$' is true (fourth row) or '$\sim A$ & $\sim B$ & C' is true (seventh row) or '$\sim A$ & $\sim B$ & $\sim C$' is true (last row). Hence the desired DBE is:

(A & B & $\sim C$) ∨ (A & $\sim B$ & $\sim C$) ∨ ($\sim A$ & $\sim B$ & C) ∨ ($\sim A$ & $\sim B$ & $\sim C$).

Now we know that any compound with n component letters requires a truth-table with 2^n rows (see Section 2.5). So clearly a DBE with n letters and 2^n disjuncts must be a tautology.

To obtain a CBE, consider first the truth-table column for '$A \vee B$'. It has an **F** in *only* the last row, and since '$A \vee B$' is a disjunction of literals in which no letter occurs more than once, it is a CBE. We can similarly produce CBE's for sentences whose columns have more than one **F**. For each **F** row, we disjoin the component letters, but this time negating those to which a **T** is assigned, and then conjoin these disjunctions. Thus there will be exactly as many conjuncts of a CBE as there are rows assigned **F**, and a CBE with n component letters and 2^n conjuncts must be a contradiction. For '$A \equiv B$', then, we look at the second and third rows, disjoining the component letters in each case and negating those to which **T** is assigned. Conjoining the two disjunctions yields the CBE '$(\sim A \vee B) \& (A \vee \sim B)$'. By the same token, the CBE for '$(A \vee B) \supset (A \& \sim C)$' is:

$$(\sim A \vee \sim B \vee \sim C) \& (\sim A \vee B \vee \sim C) \& (A \vee \sim B \vee \sim C) \& (A \vee \sim B \vee C).$$

EXERCISE

Using just **Assoc, Com, De M, Dist, Impl**, and **TE**, reduce '$(A \vee B) \supset (A \& \sim C)$' to the DBE displayed on page 243.

The symbolic notation developed in Chapter 1 for sentential logic has other applications as well. Thus far, letters like 'A' and 'B' have been taken to represent particular sentences, and variables 'p', 'q', and so on to range over those sentences. The connectives '&', 'v', and '$\sim$' have served as abbreviations for certain key uses of 'and', 'or', and 'not' respectively, and the letters **T** and **F** have been employed for the two truth-values. However, all of these expressions may be interpreted in other, highly useful ways. We shall now examine briefly such an interpretation — one which pertains to the behavior of electrical circuits. And though we shall not pursue it here, the study of these matters finds application in the design of electronic digital computers.

A two-terminal switching circuit is an array of wires and switches with a wire leading into the circuit and a wire leading out. The letters 'A', 'B', and so on will be used to represent switches, and each switch must either be *open* or *closed*. Current passes through the switch *only* when it is closed. A simple, one-switch circuit may be illustrated as follows:

(I)

Switch A is indicated by the closely spaced dots, and is in the closed position. In open position, the circuit appears like this.

Some circuits require a switch that is closed just in case another switch is open, and vice versa. In such a case, if one of the switches is labeled 'A', the other is represented as '$\sim A$'. Moreover, circuits sometimes require two or more switches to operate in tandem—opening and closing together—and thus controlled by a master switch. Those switches of a circuit linked together in such a way are all represented by the same letter. Consider then the circuit:

(II)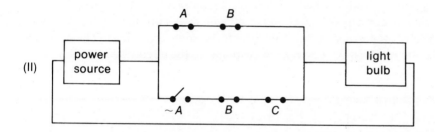

Whenever A is closed, $\sim A$ is open, and vice versa. The two switches labeled 'B', however, are always *both* open or (as above) *both* closed. Since $\sim A$ is open, current clearly cannot get to the light bulb via the lower route. But since *all* switches are closed on the other route, current can nonetheless reach the bulb.

Now a **parallel circuit** is one in which switches are placed as follows:

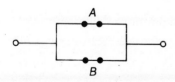

Clearly current will flow from one terminal to the other just in case *at least one* of the switches is *closed*. Since a disjunction is true just in case one disjunct is true, parallel circuits are analogues of disjunctions. The above circuit may be represented in our notation by '$A \vee B$'.

A **series circuit** is just what the name implies—two or more switches

are connected serially, as in

Here, current will flow if and only if *both* switches are closed—just as a conjunction is true if and only if both conjuncts are. This circuit may therefore be represented by '*A* & *B*'.

A **series-parallel circuit** is any circuit that is either series or parallel or a combination of the two. Circuit (II) above is such a combination; it consists of two series circuits in parallel. Current thus reaches the bulb if and only if *either A* and *B* are both closed *or* if $\sim A$, *B*, and *C* are all closed. The circuit is therefore expressible by '$(A \& B) \lor (\sim A \& B \& C)$', which incidentally is in DNF.

Given any sentence of our notation whose only connectives are '&', '$\lor$', and '$\sim$' and in which '$\sim$' attaches *only* to single letters, we can construct a corresponding circuit that will pass current under conditions exactly analogous to those on which the sentence is true. The sentence '$(A \& \sim B) \lor [B \& (C \lor A)]$' corresponds to the circuit:

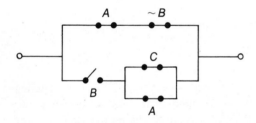

Here, current can flow from terminal to terminal, since all switches but *B* are closed. While the current cannot get through the lower series circuit, it can get through the upper. By the same token, the sentence '$(A \& \sim B) \lor [B \& (C \lor A)]$' is true when '*A*' and '*C*' are true but '*B*' is false, for on that assignment the left disjunct is true.

But now let us suppose the switches are positioned differently—and analogously, we suppose a different truth-value assignment to the component letters.

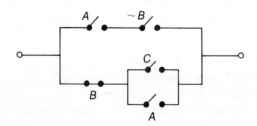

Current cannot get through the upper series, since both switches there are open; similarly, it cannot pass through the lower series, because both switches in the component parallel circuit are open. So current will not flow when switches *A* and *C* are open and *B* is closed. Likewise, '(*A* & ~*B*) v [*B* & (*C* v *A*)]' is false when '*A*' and '*C*' are false and '*B*' is true, for each component conjunction has at least one false conjunct.

It should be obvious by now that since circuits are expressible by the compound sentences of our notation, the complete behavior of a circuit for all possible positions of its switches is displayed by the truth-table for its corresponding sentence. And this means that we can *design* circuits to behave in a desired fashion by first constructing the truth-table and then finding a sentence having that column of **T**'s and **F**'s; from there, the required circuit is easily produced. As an elementary example, suppose the light in Sam's basement is connected to a switch in the kitchen and one in the living room. And further suppose that (for whatever reason) Sam wishes it to be on just in case exactly one of the switches is thrown. Where '*K*' stands for the kitchen switch and '*L*' for the living-room switch, the behavior of the required circuit is specified by the following table:

K	*L*	
T	T	F
T	F	T
F	T	T
F	F	F

(In the columns for '*K*' and '*L*', a **T** in effect means *closed*, an **F** means *open*. But in the third column, a **T** represents *current flows* while an **F** represents *current does not flow*.) To find a sentence having that arrangement of **T**'s and **F**'s, we need only produce a Boolean expansion in the way described earlier. The DBE in this case is '(*K* & ~*L*) v (~*K* & *L*)', and hence a circuit that will do the job is

(III)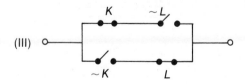

Here we have two series circuits in parallel. But suppose the layout of the house is such that two parallel circuits in series would be more convenient to install. In such case, we need only produce the CBE from our table, '($\sim K$ v $\sim L$) & (K v L). Hence another circuit that will do the *same* job is

(IV)

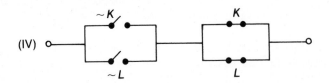

Just as '(K & $\sim L$) v ($\sim K$ & L)' and '($\sim K$ v $\sim L$) & (K v L)' are equivalent sentences, so (III) and (IV) may be described as "equivalent" circuits. The two sentences are *true* under the same conditions, and *current will flow* through the two circuits under the same conditions.

It is not always easy to find the truth-table that will lay out the behavior we desire of a circuit. But once found, it is a simple task to find the circuit, since Boolean expansions provide a mechanical technique for producing the required sentences. Suppose the Supreme Court of Transylvania consists of three judges: Allslop, Bickley, and Chief Justice Coffman. Each has a switch that when closed allows current to flow in a circuit wired to a light atop the courthouse. Each closes his switch when his verdict is "guilty" and leaves it open otherwise. The rules say that a defendant is found guilty only if *at least two* judges so decree, but that Coffman and only he retains veto power. That is, even if *both* of the others close their switches, current will not flow to the light unless Coffman closes his as well. However, in any other circumstance where at least two switches are closed, current reaches the light. To find the appropriate circuit, we first construct a truth-table using obvious letters for the three switches:

A	B	C	
T	T	T	T
T	T	F	F
T	F	T	T
T	F	F	F
F	T	T	T
F	T	F	F
F	F	T	F
F	F	F	F

The second row illustrates Coffman's veto power; current does not flow even though other justices have closed their switches. Only in the first, third, and fifth circumstances does the light go on. The corresponding

DBE, then, is:

$(A \& B \& C) \lor (A \& {\sim}B \& C) \lor ({\sim}A \& B \& C)$.

Thus a circuit that will satisfy our conditions would be:

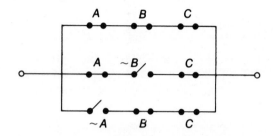

Were an electrician to install such a circuit, however, he would in all likelihood be fired. For there are many *simpler* circuits—that is, with fewer switches and wires—that are equivalent. To find such a circuit, we can try to reduce the above DBE to a simpler normal form sentence. We first commute (three times) to bring '*C*' to the left of each disjunct and then distribute, obtaining

$C \& [(A \& B) \lor (A \& {\sim}B) \lor ({\sim}A \& B)]$ Com, Dist

We now proceed to distribute the part '$(A \& B) \lor (A \& {\sim}B)$':

$C \& [(A \& (B \lor {\sim}B)) \lor ({\sim}A \& B)]$ Dist

We now obtain:

$C \& [A \lor ({\sim}A \& B)]$ TE
$C \& [(A \lor {\sim}A) \& (A \lor B)]$ Dist
$C \& (A \lor B)$ Com, TE

Thus we have our equivalent circuit.

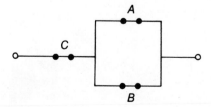

So while the DBE (or of course CBE) can always be found with no mental exertion whatever, it rarely yields the simplest circuit for the task at hand.

EXERCISES

I. Construct circuits for each of the following.
    1. A v $[B$ & $(\sim B$ v $(C$ & $\sim A))]$
    2. $(A$ v $\sim A)$ v $(B$ & $\sim B)$
    3. $\{\{[(A$ v $B)$ & $(A$ v $\sim B)$ & $B]$ v $C\}$ & $D\}$ v $(C$ & $\sim B)$

II. Design a circuit for a voting indicator that lights if and only if at least two of the voters, A, B, and C, vote *yes* or else at least one of B and C votes *yes* but A votes *no*. Can you find a circuit with just *two* switches that will do the job?

APPENDIX B

TRUTH TREES[1]

If 'A' is true, the disjunction '$A \lor B$' is also true regardless of the truth-value of 'B'. This may be put by saying that the truth of 'A' is a *sufficient* condition for the truth of '$A \lor B$'—the latter's truth follows directly from that of the former. However, '$A \lor B$' may be true even though 'A' is false, and hence the truth of 'A' is *not* a *necessary* condition for the truth of '$A \lor B$'. That is, 'A' does not *have* to be true in order for '$A \lor B$' to be true. Obviously, the same remarks apply to 'B'; its truth is a sufficient— but not a necessary—condition for the truth of '$A \lor B$'. With conjunctions, however, the situation is reversed. The truth of 'A' is not a sufficient condition for the truth of '$A \,\&\, B$' because a conjunction is true only when *both* conjuncts are. But the truth of 'A' clearly is a necessary condition for the truth of '$A \,\&\, B$' since the latter cannot be true unless 'A' is. Likewise, the truth of 'B' is necessary but not sufficient for the truth of '$A \,\&\, B$'. So, while the truth of a single disjunct is a sufficient (but not a necessary) condition for the truth of the disjunction itself, the truth of a single conjunct is a necessary (but not a sufficient) condition for the truth of the conjunction.

It will be useful here to have a means of displaying such relationships between compound sentences and their components. Where each component is such that its truth is alone sufficient for the truth of the compound, we shall "branch out" those components. The disjunction '$A \lor B$' thus gives rise to this diagram:

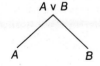

[1]The truth-tree method presented here is a version first used by Richard C. Jeffrey in his *Formal Logic: Its Scope and Limits*, New York: McGraw-Hill Company, 1967. A similar discussion may be found in Hugues Leblanc and William A. Wisdom, *Deductive Logic*, Boston: Allyn and Bacon, Inc., 1972.

The fork here indicates that the compound will be true if and only if *at least one* of the components displayed at the tips is true.

Where each component is such that its truth is a necessary condition for the truth of the compound, we shall write the components directly below the compound, one underneath the other. So for '*A* & *B*' we write:

A & *B*
  A
  B

Writing '*A*' and '*B*' in a column indicates that each *must* be true for the compound to be true (and though neither the truth of '*A*' nor that of '*B*' is alone sufficient for the truth of '*A* & *B*', their *joint* truth is of course sufficient).

We will call such diagrams — for reasons which will soon be apparent — **truth trees**, and will say that in our two examples '*A* ∨ *B*' and '*A* & *B*' have been **decomposed**, that is, broken down into sentence letters or their negations. Once the conventions concerning branching and columns are understood, it is clear that such diagrams spell out all possible ways in which the compounds may be true. The decomposition of more complex sentences requires of course larger trees.

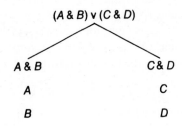

Here, the main connective is '∨', so we begin with a fork — one for each disjunct — and since both disjuncts are conjunctions we decompose each by writing its conjuncts beneath. Having done this, the truth-conditions of the compound are displayed clearly: it will be true just in case both '*A*' and '*B*' are true (left branch) or both '*C*' and '*D*' are true (right branch). That is, in the corresponding 16-row truth-table, those rows in which '*A*' and '*B*' are both assigned **T** and those in which '*C*' and '*D*' are both assigned **T** will also have a **T** assigned to '(*A* & *B*) ∨ (*C* & *D*)'. Any other assignment of truth-values to the four component letters results in a falsehood.

Let us now tackle a more complicated example, the conjunction '$(A \lor B) \& (C \lor D)$'. Since it is a conjunction, we begin by listing the conjuncts underneath:

$(A \lor B) \& (C \lor D)$
    $A \lor B$
    $C \lor D$

But since both conjuncts are themselves compound, each must be decomposed. Working on '$A \lor B$' first, we obtain:

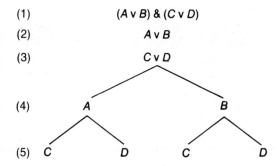

(1) $(A \lor B) \& (C \lor D)$

(2) $A \lor B$

(3) $C \lor D$

(4) A B

(5) C D C D

Here, '$A \lor B$' is decomposed at (4), and '$C \lor D$' at (5). Notice that '$C \lor D$' must be decomposed at *two* points, once for each disjunct of '$A \lor B$'. For the entire compound to be true, '$A \lor B$' must be true (as (2) tells us), but the fork between (3) and (4) shows us that for it to be true either 'A' or 'B' must be true. So both of these possibilities must be accounted for when decomposing the other disjunction if we are to delineate *all* conditions under which the compound will be true. Decomposing '$C \lor D$' at (5), then, completes our work. The resulting tree thus has four *branches*, two passing upwards from (5) through 'A' on the left, the other two through 'B' on the right. Each lays out a condition under which the original compound will be true, and together are exhaustive of such conditions. So '$(A \lor B) \& (C \lor D)$' will be true just in case:

'C' and 'A' are both true (leftmost branch)
'D' and 'A' are both true (left-middle branch)
'C' and 'B' are both true (right-middle branch)
'D' and 'B' are both true (rightmost branch).

Plainly the above tree accomplishes the same end as a 16-row truth-table but with considerably less space and effort; all conditions under which

'$(A \lor B) \,\&\, (C \lor D)$' is true have been laid out and may be discerned by tracing through each branch of the tree.

Truth trees typically afford us a quicker, more facile method of establishing both validity and invalidity than do truth-tables. They thus constitute an alternative short-cut technique to that presented in Section 2.4. We know that an argument is (truth-table) valid just in case there is no assignment of truth-values to its component letters according to which all premises come out true but the conclusion false — in other words, just in case there is no assignment on which the premises and the *negation* of the conclusion all come out true. Now truth trees provide us with an efficient means of testing a set of sentences for *consistency* — for whether there is an assignment making *all* sentences of the set true (cf. Section 2.3). So we can determine whether an argument is valid or not by constructing a corresponding truth tree for its premises and negated conclusion, and then noting whether one of its branches lays out a truth-value assignment on which all are true (in other words, on which the premises are true and the conclusion false). As an example, let us consider the argument

$(A \,\&\, B) \supset \sim C$
$D \lor (A \,\&\, B)$
$\therefore C \supset (D \lor E)$.

To construct a tree, we first list and number the premises along with the *negated* conclusion.

(1) $(A \,\&\, B) \supset \sim C$
(2) $D \lor (A \,\&\, B)$
(3) $\sim[C \supset (D \lor E)]$

Next, we pick one of these (it does not matter which we begin with) and decompose it. Beginning with (3), we note that it is a negated conditional, and since a conditional is false only when its antecedent is true and consequent false, its negation will be *true* in just such a circumstance. Thus a sentence of the form '$\sim(p \supset q)$' is true just when the sentence represented by 'p' is true and that by 'q' false (hence '$\sim q$' true). That is, the truth of 'p' and of '$\sim q$' are both necessary conditions for the whole to be true. The negation of a conditional is thus decomposed columnwise:

$\sim(p \supset q)$
p
$\sim q$

Decomposing (3), then, we obtain:

(1)   (A & B) ⊃ ~C
(2)   D v (A & B)
(3) √ ~[C ⊃ (D v E)]
(4)   C   (3)
(5)   ~(D v E)   (3)

The checkmark to the left of (3) means simply that the third sentence has been decomposed. The use of checks will enable us to keep track of which compounds have been decomposed at a given point and which have not. The citations at the right show of course that (4) and (5) were both obtained by decomposing (3). Now since (5) itself is a compound, it is decomposed before moving on to (1) and (2), for ultimately we want each of the three sentences with which we began to be broken down into sentence letters or their negations. (5) is the negation of a disjunction and so will be true just in case 'D' and 'E' are both false, that is, '~D' and '~E' are both true. Sentences of the form '~(p v q)' are hence decomposed by writing '~p' and '~q' directly underneath, Our tree now becomes:

(1)   (A & B) ⊃ ~C
(2)   D v (A & B)
(3) √ ~[C ⊃ (D v E)]
(4)   C   (3)
(5) √ ~(D v E)   (3)
(6)   ~D   (5)
(7)   ~E   (5)

We may now turn to (2). Since it is a disjunction, a fork is called for after step (7):

(1)   (A & B) ⊃ ~C
(2)   √ D v (A & B)
(3)   √ ~[C ⊃ (D v E)]
(4)   C   (3)
(5)   √ ~(D v E)   (3)
(6)   ~D   (5)
(7)   ~E   (5)

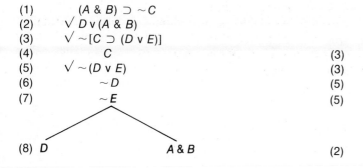

(8) D   A & B   (2)

Let us take stock of the situation here. If the checked compounds (2) and (3) are *both* true, then *either* the seven sentences from (2) down through '*D*' on the left branch are all true *or* the seven sentences from (2) down through '*A & B*' on the right branch are all true. But the first of these alternatives is impossible, since for '*D*' to be true '~*D*' at (6) must be false. That is, there is no assignment of truth-values on which *all* sentences on the left branch will be true inasmuch as '*D*' and '~*D*' would *per impossible* have to be true. So the left branch does not provide us with a condition under which (2) and (3) will both be true; in tracing the branch upwards from '*D*' at (8) we stumble into '~*D*' at (6). This being the case, we can "close off" the left branch from future consideration since we now know that *no* extension of the branch beyond '*D*' could ever correspond to a genuine truth-value assignment. We shall call any branch in which a sentence letter and its negation both occur a **closed branch**, and shall indicate such a branch by placing an '✕' beneath its last member. Thus we mark the closure of the branch from (2) through '*D*' at (8) as follows:

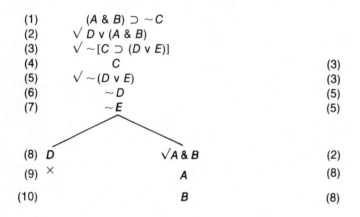

At the same time, the conjunction on the right branch has been decomposed. Note however that this branch is not closed. It is still *open* at this point since '*A & B*' has been decomposed into '*A*' and '*B*', neither of whose negations occur earlier in the branch.

Finally, we decompose (1). Since a conditional is false *only* when its antecedent is true and consequent false, it will be true if *either* the antecedent is false (hence its negation true) *or* the consequent is true. Conditionals thus require a fork

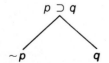

and our tree has now acquired some new growth:

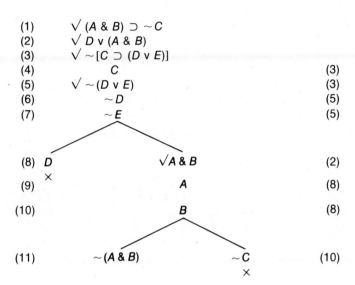

(1)	√ (A & B) ⊃ ~C	
(2)	√ D v (A & B)	
(3)	√ ~[C ⊃ (D v E)]	
(4)	C	(3)
(5)	√ ~(D v E)	(3)
(6)	~D	(5)
(7)	~E	(5)
(8)	D √A & B	(2)
(9)	× A	(8)
(10)	B	(8)
(11)	~(A & B) ~C	(10)
	×	

Another closed branch has made its appearance. The right branch at (11) cannot yield a truth-value assignment making all of (1), (2), and (3) true since '*C*' at (4) and '~*C*' at (11) would *per impossible* both have to be true. We must still however contend with the left branch. A sentence of the form '~(*p* & *q*)' will be true just in case the sentence represented by '*p*' is false ('~*p*' true) or that by '*q*' is false ('~*q*' true). We thus have

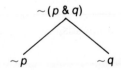

Our completed tree, then, looks like this:

(1)	√ (A & B) ⊃ ~C	
(2)	√ D v (A & B)	
(3)	√ ~[C ⊃ (D v E)]	
(4)	C	(3)
(5)	√ ~(D v E)	(3)
(6)	~D	(5)

(7)   (5)

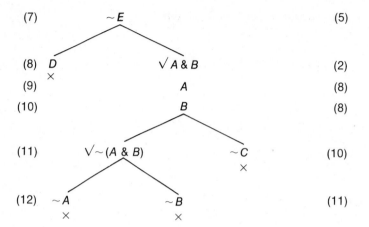

(8) *D*   √ *A & B*   (2)
       ×
(9)   *A*   (8)

(10)   *B*   (8)

(11)   √ ~(*A & B*)   ~*C*   (10)
                                     ×

(12)   ~*A*   ~*B*   (11)
        ×   ×

The new branches at (12) are both closed since '*A*' and '*B*' occur earlier in their paths. *Every* branch of the tree, then, is closed, and this means that no truth-value assignment will render all of (1), (2), and (3) true. Hence there is no assignment making the premises of our argument true and its conclusion false. The argument is therefore *valid*, and moreover has been so established with far less effort than a 32-row truth-table would have required. Let us call a tree which tests the premises and negated conclusion of an argument for joint truth a **corresponding tree** of that argument. Then we may say that for any truth-functional argument and any such tree, the argument is valid if and only if *every branch of the corresponding tree is closed*.

Rules for producing trees were introduced as they were required in the above example. We shall now collect them together along with some new ones. In all, we shall have nine rules: four governing sentences with '&', 'v', '⊃' and '≡' as main connectives, four for the negations of such sentences, and one for eliminating double curls. The justification of the rules of course resides ultimately in the truth-tables for each of these compounds.

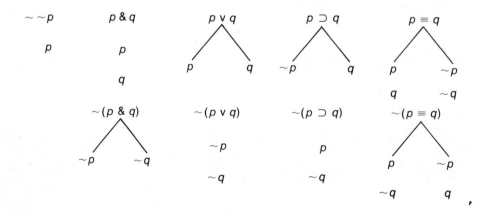

From the above remarks on validity it follows that a corresponding tree for an invalid argument will have *at least one open branch*—a branch that does not contain *both* a sentence letter and its negation. Such a branch will trace out an assignment of truth-values to the component letters of the argument which makes the premises and negated conclusion all true. As an example let us consider

$(L \& M) \supset (H \& \sim W)$
$L \lor W$
$\therefore H \lor \sim M$

Decomposing the conclusion first once again, we obtain:

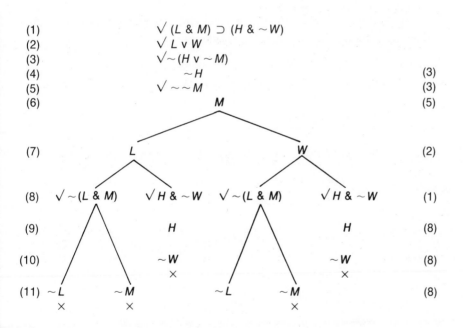

(1)	$\sqrt{}$ $(L \& M) \supset (H \& \sim W)$
(2)	$\sqrt{}$ $L \lor W$
(3)	$\sqrt{}$ $\sim (H \lor \sim M)$
(4)	$\sim H$ (3)
(5)	$\sqrt{}$ $\sim \sim M$ (3)
(6)	M (5)
(7)	L W (2)
(8)	$\sqrt{}$ $\sim (L \& M)$ $\sqrt{}$ $H \& \sim W$ $\sqrt{}$ $\sim (L \& M)$ $\sqrt{}$ $H \& \sim W$ (1)
(9)	H H (8)
(10)	$\sim W$ $\times$ $\sim W$ $\times$ (8)
(11)	$\sim L$ $\times$ $\sim M$ $\times$ $\sim L$ $\sim M$ $\times$ (8)

In steps (4)–(6), (3) is decomposed; (7) decomposes (2) and (8)–(11) do the same for (1). The third branch from the right at (11) is *open* since 'L' does not occur earlier in that branch (it does of course occur on the *left* side at (7), but for that reason is not in the upward path leading from the righthand '$\sim L$' at (11)). This branch therefore delineates an assignment of truth-values to the letters 'L', 'W', 'M', and 'H' on which all of (1), (2), and (3) come out true, and consequently establishes the argument as invalid. The assignment in question may be found by tracing through the open branch, noting whether the sentence letters in its path occur by themselves or negated. Thus, (1), (2), and (3) will all be true (hence the premises true and conclusion false) when 'L' is false (line (11)), 'W' is true (line (7)), 'M' is true (line (6)), and 'H' is false (line (4)).

Some further points about this tree are worth mentioning. First, unlike the preceding one, this tree requires a sentence to be decomposed twice, viz. premise (1) at step (8). This of course is because the branches issuing from (7) are both open at that point, and thus an exhaustive search for a truth-value assignment making all of (1)–(3) true requires that the truth-conditions of (1) be set down for each alternative at (7). Second, the double negative at (5)—like any other compound—must be checked after its decomposition. Lastly, note that the third branch from the left at (11) is closed even though 'W'—the contradictory of its very last member—does not occur earlier on the branch. It suffices for just one of the conjuncts of 'H & ~W' to have a contradictory elsewhere on the branch, and '~H' occurs at (4).

A further example illustrates some less common features a truth tree may exhibit. The validity of the argument

A ⊃ ~E
A
B & ~(C & D)
∴ E v (B & A)

is shown by a corresponding tree in which both branches are closed:

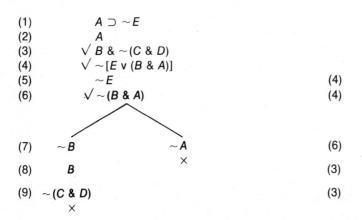

(1)   A ⊃ ~E
(2)   A
(3)   √ B & ~(C & D)
(4)   √ ~[E v (B & A)]
(5)   ~E   (4)
(6)   √ ~(B & A)   (4)

(7)   ~B   ~A   (6)
                        ×
(8)   B   (3)

(9) ~(C & D)   (3)
        ×

The right branch at (7) is closed since 'A' occurs as a premise at (2). So when a single letter or its negation occurs as a premise, it contributes to the closing of a branch just as if it had been generated through decomposition. However, since it requires no decomposition, it is *not* checked off. Notice that premise (1) also has not been checked; this is because once (3) and (4) were decomposed, all branches were closed. Thus, (1) is a superfluous premise—the conclusion following from (2) and (3) *alone.*

Further, since the right branch closes at (7), premise (3) is decomposed at steps (8) and (9) only for the left branch. Since both '*B*' and '~*B*' have now been produced, the branch is now closed and the tree is completed. Decomposition of (9) would at this stage be pointless, and so it is left unchecked. Hence, *not all* trees will have every compound checked, for as (1) and (9) illustrate, decomposition is unnecessary in certain cases. And where a premise or conclusion consists of a single letter or its negation, it will not be checked either.

The argument forms associated with such elementary rules of inference as **Simplification**, **Conjunction**, *Modus Ponens*, and **Hypothetical Syllogism** may be shown valid via truth trees (see Sections 2.3, 3.1, and 3.2):

Simplification	Conjunction	*Modus Ponens*	Hypothetical Syllogism

√ *p* & *q*

~*p*

p

q

×

—

p

q

√ ~(*p* & *q*)

~*p*   ~*q*

×   ×

—

√ *p* ⊃ *q*

p

~*q*

~*p*   *q*

×   ×

—

√ *p* ⊃ *q*

√ *q* ⊃ *r*

√ ~(*p* ⊃ *r*)

p

~*r*

~*q*   *r*

×

~*p*   *q*

×   ×

Trees for the remaining elementary rules are left as exercises for the reader. The fallaciousness of "affirming the consequent" (see Sections 2.3 and 3.2) is shown by the tree below ("denying the antecedent" is left to the reader:

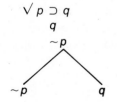

√ *p* ⊃ *q*

q

~*p*

~*p*   *q*

In each of our worked examples, the negated conclusion of the argument being tested was decomposed first. As stated earlier, however, it does not matter which sentence we begin with, and it is sometimes more convenient to decompose one of the premises first. The general rule to follow is: *decompose first those sentences requiring no forks* (sentences of the forms '$\sim\sim p$', '$p \,\&\, q$', '$\sim(p \vee q)$', and '$\sim(P \supset q)$', for the tree will be far less bulky and cumbersome if branching is kept to a minimum. Moreover, when faced with a choice of decomposing compounds both of which require forks, it is obviously wise to see if one of them will lead immediately to a closed branch. Decomposing it first will simplify later work.

The tree method is essentially a means of setting out the truth-conditions for each member of a set of sentences by systematically eliminating their connectives until only sentence letters and their negations remain. If at any point in the decomposition process *all* branches are closed, we may stop our work since the set of sentences with which we began has been shown not to admit of a truth-value assignment making all of them true. On the other hand, if at least one branch remains open after all compounds have been decomposed (whether they be in the original set or produced in the decomposition process), then there is at least one truth-value assignment on which all sentences in the set are true. To retrieve such an assignment, one need only trace upwards through the branch, assigning an **F** to each negated letter and a **T** to the rest.

EXERCISES

 I. Using the tree method, determine whether the arguments in Exercise I of Section 2.4 are valid.

 II. Using the tree method, show the validity of the argument forms corresponding to the rules: **Addition**, **Disjunctive Syllogism**, *Modus Tollens*, and **Constructive Dilemma**.

 III. Explain how the tree method may be used to determine whether or not a sentence is a *tautology*. Then determine whether any of the following are tautologies:

  1. $A \vee \sim(B \,\&\, A)$

  2. $(A \,\&\, \sim B) \vee \sim[\sim(B \,\&\, C) \,\&\, (A \,\&\, C)]$

  3. $(A \,\&\, \sim B) \,\&\, [\sim A \vee (B \,\&\, A)]$

 IV. Explain how the tree method may be used to determine whether or not a pair of sentences are *equivalent*. Then determine whether any of the following pairs are equivalent:

  1. $\sim(E \vee \sim F)$; $F \supset \sim E$

  2. $\sim P \vee \sim(R \,\&\, Q)$; $P \supset (R \supset Q)$

  3. $A \vee (B \,\&\, \sim C)$; $(A \vee \sim C) \,\&\, (A \vee B)$

APPENDIX C

MATHEMATICAL INDUCTION

When, in 1821, Sir John Cecil, first holder of the Stopes Chair of Archeology, entered the sacred tomb of an ancient pharoh, a curse was placed on him. And the story goes that when such a curse falls on an academic intruder, it falls on his successor as well. If these be facts—that a curse befell the initial occupant of the Stopes Chair, and that each occupant is such that if a curse befalls him then it befalls his successor as well—we are bound to conclude that all who have occupied the Stopes Chair are accursed.

Somewhat similar inferences are common in mathematics, where the rule of inference in question can be formulated precisely. For example:

The number 1, plus its square, is an even number. And each positive integer is such that if the sum of it and its square is even, then the same is true of its successor. For if $m + m^2$ is even then clearly $(m + m^2) + 2(m + 1)$ is also, and $(m + m^2) + 2(m + 1) = (m + m^2) + (2m + 2) = (m + 1) + (m^2 + 2m + 1) = (m + 1) + (m + 1)^2$. We are clearly entitled to conclude that the sum of any positive integer and its square is even—for we have shown that this is true of the first positive integer, and that it carries over from each to the next.

In general, if we show that the number 1 satisfies a certain condition and show also that whenever *any* positive integer satisfies that condition then the number after it satisfies that condition as well, we will be entitled to conclude that *all* positive integers satisfy the condition in question. For the second of our claims in effect ensures that each will have inherited it from its predecessor, beginning with 2's having inherited it from 1.

To formulate the general rule of inference at work here, let '...x...' represent any sequence of English words, mathematical symbols, and so on, that becomes a sentence if each occurrence of '*x*' in it is replaced by a numeral.

The Principle of Mathematical Induction.

From

...1...

and

For each positive integer *m*, if ...*m*... then ...*m*+1...

one may infer

For each positive integer *n*, ...*n*...

When one uses this principle to show that all positive integers satisfy a certain condition, his work will always divide into two parts, as did our proof that the sum of a number and its square is always even. First he will show that the number 1 satisfies the condition (this part of his work is called the **base case**); and second, he will show that if any number *m* (*m* ⩾ 1) satisfies the condition then the number after it, *m*+1, satisfies the condition also (this part is called the **induction step**). The principle of mathematical induction then justifies his inferring from these two sub-results that all positive integers satisfy the condition in question. (It's as if with the first he assigned to Edward I, and with the second to his heirs in perpetuity, the title to certain lands.)

To see how this works in practice, let's follow another example through in detail. We have seen how to show that for each positive integer *n*, $n+n^2$ is even. This means, of course, that for each such *n*, $\frac{1}{2}(n+n^2)$ will be a positive integer. Suppose we compute the first few numbers of this sort.

n	n^2	$n+n^2$	$\frac{1}{2}(n+n^2)$
1	1	2	1
2	4	6	3
3	9	12	6
4	16	20	10
5	25	30	15
.	.	.	.
.	.	.	.
.	.	.	.

The series so generated (1, 3, 6, 10, 15,...) will be familiar to anyone who has ever added numbers together in sequence:

$1 = 1$

$3 = 1+2$

$6 = 1+2+3$

$10 = 1+2+3+4$

$15 = 1+2+3+4+5$

.

.

.

So it is hard to resist *conjecturing*, at least, that

For each positive integer n, $\frac{1}{2}(n+n^2) = 1+\cdots+n$.

And once the conjecture has been formulated, a proof by mathematical induction is not hard to construct. Remember that in using induction, we must establish two things: that $\frac{1}{2}(1+1^2) = 1$; and that for each positive integer m, if $\frac{1}{2}(m+m^2) = 1+\cdots+m$ then $\frac{1}{2}((m+1)+(m+1)^2) = 1+\cdots+(m+1)$. The first of these is quite trivial:

BASE CASE $\frac{1}{2}(1+1^2) = 1$, for

$\frac{1}{2}(1+1^2) = \frac{1}{2}(1+1)$

$\qquad\qquad = 2/2$

$\qquad\qquad = 1$

The second part requires a bit more work:

INDUCTION STEP   Each positive integer m is such that if $\frac{1}{2}(m+m^2) = 1+\cdots+m$ then $\frac{1}{2}((m+1)+(m+1)^2) = 1+\cdots+(m+1)$. For let m be any positive integer and assume that $\frac{1}{2}(m+m^2) = 1+\cdots+m$. Now

$\frac{1}{2}((m+1)+(m+1)^2) = \frac{1}{2}((m+1)+(m^2+2m+1))$

$\qquad\qquad = \frac{1}{2}(m+m^2+2m+2)$

$\qquad\qquad = \frac{1}{2}(m+m^2)+\frac{1}{2}(2m+2)$

$\qquad\qquad = \frac{1}{2}(m+m^2)+(m+1)$

So $\frac{1}{2}((m+1)+(m+1)^2) = \frac{1}{2}(m+m^2)+(m+1)$. But we know, from our assumption, that $\frac{1}{2}(m+m^2) = 1+\cdots+m$. Consequently $\frac{1}{2}((m+1)+(m+1)^2) = 1+\cdots+m+(m+1)$, completing our proof.

Beyond its obvious usefulness in establishing generalizations concerning numbers, mathematical induction can often be employed in establishing general results in logic. In Section 4.2, for example, we use it (tacitly) to show that for each positive integer n, any line of any deduction whose line-number is no greater than n must have a **T** in each row of its truth-table in which all the premises, $\mathscr{P}_1, \ldots, \mathscr{P}_k$, of that deduction have **T**'s. For the base case, we show that this is true of all lines of deductions from $\mathscr{P}_1, \ldots, \mathscr{P}_k$ whose line-numbers are no greater than 1; and for the induction step we show that if all lines of all deductions from $\mathscr{P}_1, \ldots, \mathscr{P}_k$ whose line-numbers are no greater than m (m any positive integer) have **T**'s when $\mathscr{P}_1$ through $\mathscr{P}_k$ do, then the same holds of all lines of such deductions whose line-numbers are no greater than $m + 1$.

As another example, let us establish Chapter 2's claim that

For each positive integer n, there are exactly 2^n ways of assigning truth-values **T** and **F** to n distinct sentence letters.

For the base case we must show that there are exactly 2^1 (that is, 2) ways of assigning truth-values to 1 sentence letter, and this is clear enough: we can assign a **T** to it or an **F**, but there is no third possibility.

For the induction step, let m be any positive integer and assume that there are exactly 2^m ways of assigning truth-values **T** and **F** to m distinct sentence letters. Suppose, now, we have $m + 1$ sentence letters, say 'A_1', ..., 'A_m', 'A_{m+1}'. By hypothesis there are exactly 2^m ways of assigning **T** and **F** to 'A_1', ..., 'A_m'. And for *each* of these 2^m assignments we have two choices of assignment to 'A_{m+1}': we may assign it a **T** or we may assign it an **F**. In all, then, there are 2×2^m ways of assigning truth-values to the $m + 1$ sentence letters in question; and since $2 \times 2^m = 2^{m+1}$, our induction is complete.

EXERCISES

1. Where n is any positive integer, $n + n^2 = n(n + 1)$. Show, by induction, that

    For each positive integer n, either n or $n + 1$ is even,

    thus providing an alternate proof that $n + n^2$ is always even.

2. Prove by induction that for each positive integer n, $1^2 + \cdots + n^2 = \frac{1}{6}(n(n + 1)(2n + 1))$.

3. Prove that for each positive integer n, $1^3 + \cdots + n^3 = (1 + \cdots + n)^2$.

4. Prove that for each positive integer n, there are exactly 2^{2^n} distinct truth-functions (see Section 2.5) of n arguments.

5. Without using any results from Chapter 4, prove that every theorem of the deductive system developed in Section 3.1 is a tautology (in

the sense of Section 2.1) by using induction to establish

For each positive integer n, each line of each n-line proof is a tautology.

*6. The principle formulated here is sometimes called the **Principle of Weak Mathematical Induction**, with the **Principle of Strong Mathematical Induction** being formulated as follows (where '---x---' represents any sequence of English words, mathematical symbols, and so on, that becomes a sentence if each occurrence of 'x' in it is replaced by a numeral):

From

--- 1 ---

and

For each positive integer m, if ---k--- for each positive integer k such that $k \leq m$ then ---$m+1$---

one may infer

For each positive integer n, ---n---

Suppose we have established --- 1 --- and also that for each positive integer m, if ---k--- for each positive integer k such that $k \leq m$ then ---$m+1$---. Do we need the principle of *strong* induction to conclude that ---n--- for each positive integer n, or does weak induction suffice if we choose our substitution for '...n...' carefully?

APPENDIX D

AXIOMATIC SYSTEMS
A Preface to Their Study

Any collection of signs, together with a collection of sequences of those signs, constitutes what we shall here regard as a **formal language**. We call the former the **primitive symbols** of the language in question, and the latter its **well-formed formulas** ('wffs', for short).

EXAMPLE The Formal Language *N*.

Primitive Symbols: '|' (the 'stroke') and '–' (the 'dash').

Wffs: those finite sequences of such symbols that consist of one or more strokes, followed by a single dash, followed by one or more strokes.

Note that in the specification of a formal language, no "meanings" are assigned to the primitive symbols.

Of course, full-blooded natural languages such as English involve more than just "alphabets" (primitive symbols) and "sentences" (wffs), but the more anemic *formal* languages serve as adequate bases for the systems we are interested in constructing here.

Given such a formal language, we may select certain of its wffs as **axioms**, and lay down certain **rules of inference** (or **transformation rules**) permitting passage from some of its wffs to others. And the resulting package, consisting of the formal language, together with the axioms and the rules, is an **axiomatic system**.

EXAMPLE The Axiomatic System **N**.

Formal Language: *N*

Axiom: |–|

Rule: From any wff

$$\mathcal{S} - \mathcal{T}$$

one may pass to the corresponding wff

$$|\mathcal{S} - \mathcal{T}|$$

In discussing such a system we make use not of a formal language, but of a natural language, ordinary English. Since the latter is used for talking *about* the former, we call it our **metalanguage**. In addition to ordinary English expressions, our metalanguage contains **metalinguistic variables**, '$\mathcal{S}$', '$\mathcal{T}$', and so on, that is, variables which range over sequences of primitive symbols of the system being discussed. Thus, in our statement of **N**'s rule, '$\mathcal{S} - \mathcal{T}$' is not itself a wff of **N**, but rather an expression (in the metalanguage) used for talking about wffs of **N** — in the present case, those sequences of primitive symbols consisting of any number of strokes followed by a dash followed by any number of strokes.

A first application of the rule of our sample system **N** to its axiom gives us '$||-||$'; another application of the rule, this time to the result '$||-||$' of the first, provides '$|||-|||$'; and further applications will deliver still more wffs. These wffs, generated from the axiom by successive applications of the rule, are the *theorems* of our system, and the sequence of wffs involved in the generation of each is a *proof* of it, within that system. Thus such sequences as

```
|-|      |-|      |-|          |-|
         ||-||    ||-||        ||-||
                  |||-|||       |||-|||
                               ||||-||||
```

are proofs in the system **N**, and the last line of each is one of **N**'s theorems. These notions may be defined for axiomatic systems generally: where $\mathcal{U}$ is any axiomatic system, a **proof** in $\mathcal{U}$ is a sequence of wffs of $\mathcal{U}$ in which each wff is either an axiom of $\mathcal{U}$ or the result of applying one of $\mathcal{U}$'s rules to one or more earlier wffs in the sequence; and a **theorem** of $\mathcal{U}$ is any wff which is the last line of some proof in $\mathcal{U}$.

To show that a wff *is* a theorem of a given system, then, it suffices to construct a proof of it, that is, a proof of which it is the last line. But how are we to show that certain wffs are *not* provable in particular systems? How, for example, could we show that '$|-||||$' and '$||||-||$' are *not* theorems of **N**?

A method that is often useful is provided by the following general result:

(R) If (i) every axiom of an axiomatic system $\mathscr{U}$ has some feature, and (ii) the rules of $\mathscr{U}$ "preserve" that feature in the sense that they can only lead from wffs with the feature to others that also have it, then every theorem of $\mathscr{U}$ has the feature in question.

To establish this result, notice that every theorem of an axiomatic system $\mathscr{U}$ is the last line of some proof in $\mathscr{U}$. It will do, then, to show that every line of every proof in $\mathscr{U}$ must have the feature in question, and this is easily established by mathematical induction.[1] By (i) the first lines of all proofs in $\mathscr{U}$ have the feature. And by (ii) whenever all lines of all m-line proofs have the feature so do all lines of all $(m+1)$-line proofs—for the $(m+1)$st lines of such proofs must either be axioms themselves, and so have the feature by (i), or else follow from earlier lines by one of $\mathscr{U}$'s rules, and so have the feature by (ii).

Now we can return to one of our earlier questions and, with the aid of (R), show that '|–||||' is not a theorem of **N**.

One of the features possessed by **N**'s axiom is that of *being composed of an even number of strokes*. And **N**'s rule clearly preserves this feature, for it lets us add to a wff precisely two strokes (one at each end). The general result (R) assures us, then, that all theorems of **N** have this feature—all theorems of **N** are composed of an even number of strokes. So those wffs of **N** composed of an *odd* number of strokes ('||–|||||', '||||–|||||', and so on) are not provable in **N**. In particular, '|–||||', being composed of *five* strokes, cannot be a theorem of **N**.

But what of '||||–||'? Since it contains an even number of strokes, the preceding considerations do not suffice to establish its unprovability in **N**. To do that, we shall have to find a *different* feature, one which (R) can be used to show is possessed by all theorems of **N** but which '||||–||' lacks.

To describe one such feature it is convenient to provide an **interpretation** for the wffs of **N**, that is, to assign "meanings" to them by the assignment of meanings to certain sequences of **N**'s primitive symbols. Temporarily, let us agree to regard a sequence of m strokes ($m = 1, 2, 3, \ldots$) as standing for the mth positive integer, and read '–' as 'is identical with'. Then **N**'s axiom has the feature of *being true on this interpretation*, since it is true that the number one is identical with the number one. This

---

[1] See Appendix C.

feature is also preserved by **N**'s rule, for if it is true that the number m is identical with the number n, then it is true that $m+1$ is identical with $n+1$. Of course this feature, since it involves a specific interpretation of **N**'s wffs, is considerably more complicated than the one we looked at before. But the applicability of (R) is not blocked by complexity: all theorems of **N** must have this feature, that is, all theorems of **N** must be true on the interpretation suggested. However '||||–||' is *not* true on that interpretation, since four is not identical with two, so '||||–||' cannot be a theorem of **N**.

Indeed, *no* wff of **N** in which the dash is flanked by a different number of strokes on the left than on the right is provable in **N**, for each such wff is false on the interpretation described. On the other hand, any wff *true* on this interpretation — any wff composed, for some positive integer m, of m strokes, followed by a dash, followed by m strokes again — is a theorem of **N**, since each such wff is obtainable from **N**'s axiom by $m-1$ applications of **N**'s rule. Our interpretation thus *characterizes* the theorems of **N**, for the latter are precisely the wffs of **N** true on that interpretation.

Consequently, **N** is **decidable** in the sense that there exists a mechanical procedure for determining of an arbitrary wff of **N** whether or not it is provable in **N**. One has only to count strokes, and see if the same number occur on the left as on the right.

EXERCISES

1. Consider another interpretation of the wffs of **N**, one according to which sequences of strokes are understood as before but '–' is read 'is less than or equal to'. Are all the theorems of **N** true on this interpretation? Explain.
2. Are all the wffs of **N** that are true on the interpretation described in the preceding exercise also theorems of **N**? Explain.
3. Would the answers to these questions change if we altered the interpretation so that '–' is read 'is less than'? If '–' is read 'is the sole prime factor of'? If '–' is read 'is the square of'? If '–' is read 'is an even number identical with'? (In each case, say why or why not.)
4. Construct a new axiomatic system, one just like **N** except for having additional axioms and/or rules, whose theorems can be characterized as being precisely those wffs true on the interpretation suggested in Exercise 1.

Of course the system **N** is too simple to be of much value except as an example. But more significant results are obtainable, and more interesting notions emerge, when we turn instead to

The Axiomatic System D.

Primitive Symbols: '~', '&', 'v', '(', ')', 'A', 'B', 'C',…, 'Y', 'Z', 'A_1', 'A_2',…

Wffs: (i) Each letter ('A', 'B', etc.) is a wff;
  (ii) the result of prefixing '~' to a wff is a wff;
  (iii) the result of placing '&' or 'v' between any two wffs, enclosing the whole in parentheses, is a wff;
  nothing is a wff unless it can be shown to be so on the basis of (i)–(iii) above.

Axioms: Provided $\mathscr{P}$ is a wff,

$$(\mathscr{P} \text{ v} \sim \mathscr{P})$$

is an axiom.[2]

Rules: **Conj**, **Simp**, **Add**, **DS**, **Com**, **Assoc**, **Dist**, **De M**, and **DN**.

D is, for all intents and purposes, the same system as the one developed in Section 1 of Chapter 3. Its presentation here differs from our earlier presentation in just one minor detail: the so-called "rule" **E-M I** has been dropped, the wffs it supplied being taken instead as axioms.

D's formal language was introduced in Chapter 1, and the *standard* interpretation (by means of truth-tables) of that language was treated in detail in Chapter 2. In Section 1 of Chapter 3 we established a number of **D**'s theorems, and in Chapter 4 obtained some metatheorems relating **D** to its interpretation. When, for example, the letters 'A', 'B', 'C', and so on, are taken to stand for particular English sentences, the theorems of **D** are not just true but *tautologies*, a fact which is sometimes put by saying that **D** is **sound** (less frequently, **analytic**) with respect to the interpretation in question.[3] The converse, we saw, is true as well: **D** is **complete** (or **weakly complete**) with respect to the interpretation of Chapter 2 in the sense that each wff of **D** which is a tautology is also a theorem of **D**. And these two results concerning **D**'s soundness and completeness together establish the *decidability* of **D**, truth-tables providing a mechanical procedure for determining which wffs of **D** are among its theorems and which are not.

In the case of **D**, and many other axiomatic systems as well, one is

---

[2] In chapters 2 and 3 we adopted the convention of omitting outer parentheses in an expression standing alone. Here, however, we adhere strictly to the above definition of a wff and retain all parentheses.

[3] The general result (R) obtained above makes it especially evident that **D** is sound: **D**'s axioms are all tautologies and its rules preserve tautologousness, so by (R) all theorems of **D** are tautologies. Compare Exercise 5, pp. 266–267.

interested not just in theorems and proofs, but in *deductions*. A wff $\mathscr{C}$ is **deducible from** wffs $\mathscr{P}_1, \ldots, \mathscr{P}_k$ in an axiomatic system $\mathscr{U}$ if and only if there is a finite sequence of wffs of $\mathscr{U}$ beginning with $\mathscr{P}_1, \ldots, \mathscr{P}_k$ and ending with $\mathscr{C}$ in which each intermediate wff is either one of $\mathscr{U}$'s axioms or the result of applying one of $\mathscr{U}$'s rules to earlier wffs in the sequence. (The sequence of wffs in question, of course, is called a **deduction in** $\mathscr{U}$ **of** $\mathscr{C}$ **from** $\mathscr{P}_1, \ldots, \mathscr{P}_k$.) In Chapter 3 we were primarily interested in constructing deductions in **D**, and in Chapter 4 were able to establish *strong* completeness and soundness results complementing the weaker ones: where $\mathscr{P}_1, \ldots, \mathscr{P}_k$ and $\mathscr{C}$ are wffs of **D**, if there is no assignment of truth-values to letters according to which $\mathscr{P}_1, \ldots, \mathscr{P}_k$ are assigned **T**'s but $\mathscr{C}$ is assigned an **F**, then $\mathscr{C}$ is deducible in **D** from $\mathscr{P}_1, \ldots, \mathscr{P}_k$; and conversely, if $\mathscr{C}$ is deducible in **D** from $\mathscr{P}_1, \ldots, \mathscr{P}_k$, then there is no assignment of truth-values according to which $\mathscr{P}_1, \ldots, \mathscr{P}_k$ are assigned **T**'s but $\mathscr{C}$ an **F**.

It is also of interest to ask whether there exists a theorem of **D** whose negation is provable as well, saying that **D** is **inconsistent with respect to** '$\sim$' if there is such a wff and **consistent with respect to** '$\sim$' if there is not. In light of our earlier work, the question is easily answered: if there were such a wff then both it and its negation would have to be tautologies, for all theorems of **D** are; and it is clearly impossible for the negation of any tautology to be a tautology itself.

More generally, a system is said to be **absolutely inconsistent** if *all* its wffs are theorems, and **absolutely consistent** otherwise, that is, so long as at least one of its wffs is unprovable. The appropriateness of these definitions is best brought out by noticing that they coincide with the earlier ones (tied to '$\sim$') in the case of a number of systems like **D**.[4]

(C) Let $\mathscr{U}$ be any axiomatic system whose wffs include those of **D** and whose rules include **Add** and **DS**. Then $\mathscr{U}$ is absolutely inconsistent if and only if it is inconsistent with respect to '$\sim$' (and so, is absolutely consistent if and only if consistent with respect to '$\sim$').

Half of this result is obvious: if $\mathscr{U}$ is absolutely inconsistent then *all* of its wffs are theorems and so in particular both 'A' and '$\sim A$' are, whence $\mathscr{U}$ is inconsistent with respect to '$\sim$'.

For the other half, assume that $\mathscr{U}$ is inconsistent with respect to '$\sim$'. Then there is at least one wff $\mathscr{P}$ such that both $\mathscr{P}$ and its negation are theorems of $\mathscr{U}$. Clearly, then, a proof of any wff $\mathscr{Q}$ can be constructed in $\mathscr{U}$; one need merely append the proof of $\mathscr{P}$ to the proof of its negation, and

---

[4]The *usefulness* of the notions of absolute consistency and inconsistency lies, of course, in the fact that they are applicable even to systems such as **N** which contain no primitive symbol '$\sim$'.

follow with appropriate applications of **Add** and **DS**:

$$.$$
$$.$$
$$.$$
$\sim\mathscr{P}$
$$.$$
$$.$$
$$.$$
$\mathscr{P}$
$\mathscr{P} \vee \mathscr{Q}$ [Add]
$\mathscr{Q}$ [DS]

If any such system were inconsistent with respect to '$\sim$', then, all of its wffs would be provable and so it would be complete, in a totally trivial way, with respect to any interpretation.

Of course it is possible to construct axiomatic systems equivalent in deductive power to **D** which make do with considerably fewer axioms. Let us say that a wff $\mathscr{P}^*$ of **D** is a **substitution instance** of a wff $\mathscr{P}$ of **D** just in case $\mathscr{P}^*$ can be derived from $\mathscr{P}$ by substituting for each letter in $\mathscr{P}$ some wff of **D** (the same wff replacing the same letter throughout). Recall from Chapter 2 that if $\mathscr{P}$ is a tautology ('$(A \vee \sim A)$', say) then the associated sentence form ('$(p \vee \sim p)$', in the case of this example) must be tautologous as well, so that any other wff of the same form as $\mathscr{P}$ must also be a tautology. Now each substitution instance $\mathscr{P}^*$ of any wff $\mathscr{P}$ is of the same form as $\mathscr{P}$. Consequently, *every substitution instance of a tautology is itself a tautology*. And since the theorems of **D** are (according to the soundness and completeness results obtained earlier) precisely those wffs of **D** that are tautologies, it follows that *every substitution instance of a theorem of* **D** *is itself a theorem of* **D**. The following rule of inference, then, can be added to **D** without altering its stock of theorems:

Substitution (Subs). From any theorem $\mathscr{P}$, one may infer any substitution instance $\mathscr{P}^*$ of $\mathscr{P}$.

And when that addition is made, most of **D**'s axioms become obtainable from others via the new rule, and so all but one of them can be dropped:

The Axiomatic System **D***.

Formal Language: That of **D**

Axiom: $(A \vee \sim A)$

Rules: Those of **D**, plus **Subs**

Since the missing axioms of **D** can all be recovered from **D***'s axiom by **Subs**, all theorems of **D** are theorems of **D***; and since **Subs** brings no new theorems with it, all theorems of **D*** are theorems of **D**. For similar reasons, a wff $\mathscr{C}$ is deducible in **D** from wffs $\mathscr{P}_1, \ldots, \mathscr{P}_k$ if and only if it is deducible in **D*** from those same premises. As a result, **D*** has all the features we've shown **D** to have—it is sound and complete, in both weak and strong senses, decidable, consistent with respect to '~', and absolutely consistent.

D* has an additional feature as well: it is **absolutely complete** in the sense that were any nontheorem of **D*** taken as additional axiom, the resulting system would be absolutely inconsistent. For let $\mathscr{P}$ be any wff not provable in **D***; then $\mathscr{P}$ cannot be a tautology, since all tautologies are theorems. There exists, then, an assignment of **T**'s and **F**'s to the letters occurring in $\mathscr{P}$ according to which $\mathscr{P}$ itself takes the value **F**. Let $\mathscr{P}*$ be the wff that results from $\mathscr{P}$ by substituting the wff '$(A \vee {\sim}A)$' for each letter in $\mathscr{P}$ assigned a **T** and the wff '$(A \,\&\, {\sim}A)$' for each letter in $\mathscr{P}$ assigned an **F**. Then $\mathscr{P}*$ receives the value **F** for *all* assignments of values to 'A', so $\mathscr{P}*$ is a contradiction. The negation of $\mathscr{P}*$, then, is a tautology, and is consequently a theorem of **D***. But $\mathscr{P}*$ is a substitution instance of $\mathscr{P}$, and so is obtainable (by **Subs**) in any system obtained from **D*** by adding $\mathscr{P}$ as new axiom. Any such system, then, must be inconsistent with respect to '~', since both $\mathscr{P}*$ and its negation are provable in it. And by our earlier result (C) that system must then be absolutely inconsistent as well.

By introducing the rule **Subs**, then, we can pare down **D**'s infinitely many axioms to the single axiom of **D***. And by complicating the latter, we can make do with fewer rules. If **D***'s axiom is replaced with

$$({\sim}({\sim}({\sim}A \vee B) \vee (C \vee (D \vee E))) \vee ({\sim}({\sim}D \vee A) \vee (C \vee (E \vee A)))).$$

for example, we need only **De M**, **Subs**, and the rule permitting passage from a wff $\mathscr{P}$ and a disjunction

$$({\sim}\mathscr{P} \vee \mathscr{Q})$$

to $\mathscr{Q}$. With that single axiom and those three rules, we can prove all tautologies and deduce from any wffs $\mathscr{P}_1, \ldots, \mathscr{P}_k$ all wffs deducible from them in **D***.[5]

---

[5]Compare C. A. Meredith, "Single axioms for the systems (C, N), (C, O) and (A, N) of the two-valued propositional calculus", *Journal of Computing Systems*, vol. 1 (1953), pp. 155–164.

Of course, our treatment of '&', 'v' and '~' can be extended to '⊃'. If the latter is added to our list of primitive symbols, and the characterization of our wffs is altered in the obvious way, addition of the rule **Impl** suffices to allow us to obtain extended soundness and completeness results (relative to the usual truth-table interpretation of '⊃'). Alternative axiomatizations have also been found, of course. It is possible to make do with just **Subs** and **MP** as rules, for example, so long as our axioms include:[6]

$(((A \supset B) \supset C) \supset ((C \supset A) \supset (D \supset A)))$

$((A \& B) \supset A)$

$((A \& B) \supset B)$

$(A \supset (B \supset (A \& B)))$

$((\sim A \supset \sim B) \supset (B \supset A))$

$(A \supset (A \vee B))$

$(B \supset (A \vee B))$

$((A \supset C) \supset ((B \supset C) \supset ((A \vee B) \supset C)))$

We leave it to the reader to pursue these and other matters in more advanced texts devoted to the study of axiomatic systems. Ultimately, of course, the apparatus of predicate letters, constants, variables, and quantifiers is added to the formal language of our system **D**, versions of **UI**, **UG** and the like are incorporated into the axioms and rules, and all of quantificational logic is developed axiomatically. Though we stop short of this development here, it ought to be noted that some of the most interesting results in modern logic lie in this area. Soundness and completeness results, relative to the kind of interpretation introduced briefly in Section 3 of Chapter 5 above, are obtainable. And quantificational logic, unlike the sentential calculi discussed in this Appendix, turns out to be *undecidable*—no mechanical methods for determining which wffs are provable, nor which wffs are deducible from which others, can be devised.[7]

---

[6]Compare A. N. Prior, *Formal Logic*, Oxford: Clarendon Press, 1962, pp. 302–303. If '≡' is added to our formal language, it suffices to add '$((A \equiv B) \supset ((A \supset B) \& (B \supset A)))$' and '$(((A \supset B) \& (B \supset A)) \supset (A \equiv B))$' to our axioms.

[7]For the axiomatic development of quantificational logic, and proofs of these and other results, see E. Mendelson, *Introduction to Mathematical Logic*, Princeton: Van Nostrand, 1964, or G. Hunter, *Metalogic*, Berkeley: University of California, 1971.

INDEX